Charles Darwin

CHARLES DARWIN
THE MAN AND HIS INFLUENCE

PETER J. BOWLER

UNIVERSITY PRESS

Published by the Press Syndicate of the University of Cambridge
The Pitt Building, Trumpington Street, Cambridge CB2 1RP
40 West 20th Street, New York, NY 10011-4211, USA
10 Stamford Road, Oakleigh, Melbourne 3166, Australia

First published 1990 by Blackwell Publishers Oxford
Reissued by Cambridge University Press 1996

Printed in Great Britain by Biddles Ltd, Guildford & King's Lynn

British Library Cataloguing in Publication Data available

Library of Congress Cataloging-in-Publication Data available

ISBN 0 521 56222 8 hardback
ISBN 0 521 56668 1 paperback

Contents

CONTENTS

General Editor's Preface

OUR SOCIETY depends upon science, and yet to many of us what scientists do is a mystery. The sciences are not just collections of facts but are ordered by theory; which is why Einstein could say that science was a free creation of the human mind. Though it is sometimes presented dispassionately and impersonally, science is a fully human activity; and the personalities of those who practise it are important in its progress, and often interesting to us. Looking at the lives of scientists is a way of bringing science to life.

Those scientists who appear in this series will be chosen for their eminence, but the aim of their biographers is to place them in their context. The books will be long enough for authors to write about the times as well as the life of their subjects. Science has not long been a profession, and for many eminent practitioners of the past it was very much a part-time activity: their *Lives* will therefore show them practising medicine or law, fighting wars, looking after estates or parishes, and will not simply focus upon their hours in the laboratory. How somebody earned a living, made a career and got on with family and friends is an essential part of a biography – though in this series it is the subjects' commitment to science that has got them in, and must always be at the back of the biographer's mind.

Charles Darwin is one of the most famous scientists who have ever lived, and although there have been numerous biographies and studies of his work a new *Life* needs no excuse. There is now a large Darwin industry; it is only in recent years that we have had scholarly editions of his aborted big book *Natural Selection* and his notebooks, and his full correspondence is being published in an exemplary edition. We know more about him than we used to do; and Peter Bowler is one of those who has written very illuminatingly on the background to evolutionary theory and on the reception of Darwin's ideas in his own time and in the years after his death.

Everybody connects Darwin with the theory that all organisms are descended from a common ancestor, and many know that he was by no means the first to say so. His own grandfather, J. B. Lamarck and Robert Chambers are among the celebrated 'precursors' who published evolutionary ideas long before Darwin did. What was original was that he made the idea scientifically respectable by proposing the mechanism of natural selection – not in his case as the only mechanism for evolution, for he allowed what we would call 'Lamarckian' factors a role also, but as the chief mechanism. Whereas evolution had previously been an untestable theory, closely linked with the idea of progress, Darwin made it a very powerful idea which linked together various previously distinct sciences and led to a powerful research programme. He duly found that while many of his contemporaries saw it as speculative and unnecessary, the young all took it up.

However, most of them made Darwinism more progressive than Darwin had been himself. The male barnacles which have 'degenerated' into a stomach and sex organs living within the female are as much the product of evolution by natural selection as mankind is. The process is not one which leads to better and better, or higher, organisms, but to organisms that fit better in a niche: and families can go down as well as up in the world. The idea was shockingly open ended to nineteenth-century thinkers, and indeed to many since; but a century after his death, we are probably better Darwinians than the first generation of his followers were.

This is in some ways surprising because Darwin and his writings seem so accessible. The *Origin of Species* was bought, and is bought, by many people who are not accomplished biologists and yet who find that it makes fascinating reading. So much knowledge is packed into it and Darwin's other books that they are deceptively easy. Darwin's 'gentlemanlike' manner, and his life at Down where he pottered in the garden rather like the parson–naturalist he nearly became, may make us underrate him. His contemporaries wrestling with thermodynamics are clearly geniuses, whereas in reading his books and letters we can imagine ourselves as more like Darwin. But if we stop to think about it, we find that commitment to a great idea, with a determination to test and refine it, is characteristic of great scientists but not of us ordinary busy people.

Peter Bowler's *Life* brings Darwin before us, but also shows us how he reacted to contemporary ideas (at a time when science and humanities were certainly not 'two cultures') and how his ideas were received and adapted. It is an excellent essay in cultural history, dealing with a very great man; we are lucky to be able to read it and enjoy it.

David Knight
University of Durham

Preface

As explained in chapter 1, this is not a biography of Darwin in the conventional sense. Although I have worked on the history of evolutionism for many years, I have never specialized in the study of Darwin himself and I cannot hope to compete with members of the 'Darwin industry' on their own terms. The availability of Darwin's papers has created the possibility of a new in-depth biography and I sincerely hope that such a biography will soon be produced. But instead of trying to catch up with those scholars who have a far more detailed knowledge of Darwin's writings than I shall ever have, I have chosen another tack. A really detailed biography of the kind that is now becoming possible must necessarily be an enormous volume beyond the reach of many non-specialists. In the meantime, I think there is room for a book with a different approach – a biography that takes account of the detailed scholarly work now available, yet is written at a level that will allow ordinary readers and students to gain some appreciation of the problems that confront specialist historians.

If I am not an expert on Darwin himself, I have at least devoted a significant fraction of my life to trying to understand the impact of his theory. The Darwin industry itself has overturned many of the myths surrounding his work, but there is another line of research that has also helped to throw a different light on the impact of evolutionism. Historians are beginning to suspect that the conventional image of Darwin's influence on the development of modern thought is significantly distorted. Darwin was not the only person to explore the concept of evolution in the early nineteenth century, nor did his theory of natural selection dominate the evolutionary thought of late Victorian times. Darwin's role in the development of evolutionism thus needs to be reinterpreted in the light of modern historical research. This book uses a survey of his life and work as a vehicle for encouraging the reader to think more carefully about how we should interpret a man whose ideas have been used and misused in so many different ways, both inside and outside science.

Peter J. Bowler
Belfast

1

The Problem of Interpretation

N O O N E D O U B T S that Charles Darwin played a major
role in the development of modern science and – thanks
to the controversial nature of his theory – in the growth of modern
attitudes and values. But more than any other great scientist, Darwin
remains surrounded in controversy. For every biologist who hails him
as the founder of modern evolution theory, there is an anti-Darwinian
who claims that he led modern thought up a blind alley and thereby
undermined the traditional values of Western civilization. There are
even a few scientists who believe that modern, i.e. Darwinian, evolution
theory has taken a wrong turn. Outside science, religious and political
thinkers of widely differing backgrounds will agree on at least one
thing: that the materialism of the Darwinian theory paved the way for
the emergence of a ruthless 'social Darwinism' in which all moral values
have been abrogated. How, then, do we approach the life of a man
whose work still arouses such strong and contradictory feelings? It is
clearly impossible to launch directly into a minute study of his career.
If the events in Darwin's life are to have any meaning for us, we must
first uncover at least some of the reasons why his theory evoked and
continues to evoke these attitudes. Darwin has become an almost
mythological figure in the emergence of modern culture, his achieve-

ments overlaid with layers of interpretation and misinterpretation designed to influence our judgement of what he did. The first step in any biography must be to untangle the complex motivations that have led people to their various evaluations of the Darwinian theory.

The Man and the Myths

Even in his own lifetime, the name of Charles Darwin had become a multi-faceted symbol used by different people for their own different purposes. To rationalists he epitomized the scientists' ability to penetrate areas of knowledge once obscured by religious dogma. For liberals his theory underpinned an optimistic philosophy of progress, providing a conceptual guarantee that things would continue to improve if only nature and society were left to develop on their own. Conservative theologians protested that his theory undermined moral values and the social order by reducing the human race to the level of animals. Extremists on both sides saw the debate over the *Origin of Species* as a battle in the ongoing war between science and religion for control of the human mind.

Yet the dust of battle settled remarkably quickly. When Darwin died in 1882 he was buried in Westminster Abbey, ostensibly as a national hero of scientific discovery. Yet there can be little doubt that in accepting him as a hero of scientific discovery the Victorians symbolized a major transformation of their culture.[1] For Darwin to have been accorded this honour, the fears of those conservative thinkers who saw his theory as a threat to the very foundations of traditional religion and morality had to be swept aside, if only temporarily. A truce must have been declared in the war between science and religion, allowing evolutionism to be uncoupled from its materialistic implications. Evolution had, in fact, become a symbol of the Victorians' faith in their own ability to participate in the inevitable ascent of the universe towards perfection.

Today Darwin's name continues to evoke a series of powerful but mutually contradictory images in our minds. Many scientists believe

PLATE 1 Darwin as a public figure: caricature from the periodical
Vanity Fair,
September 30th, 1871

that his theory of evolution by natural selection has been triumphantly vindicated, providing the basis for the modern attempt to understand the development of life on earth. With the encouragement of these modern Darwinians, historians of science have mounted an unprecedented campaign to reconstruct the process by which Darwin made his discovery. So many scholars are now involved in the collection, publication and analysis of Darwin's private papers that we routinely speak of a 'Darwin industry'. This activity is clearly based on the assumption that the discovery of natural selection marks a turning point in the development of modern science. Yet even within the ranks of professional biologists there are a few who think that Darwin's ideas were misguided and will have to be modified or even abandoned if science is to advance much further. For an anti-Darwinian biologist such as Soren Lovtrup, the story of how natural selection was discovered and popularized is more problematic: he asks how it was possible for generations of biologists to have been led so far along the wrong track.[2] Even from the perspective of a specialized historian of science, evaluation of Darwin's achievements thus requires a value judgement to be made from the very beginning.

Outside science, Darwin's name continues to be used as a symbol for people's religious, philosophical and ideological perspectives. Those scientists who hail Darwin as a hero of discovery are themselves using him to make a point about the value of science, rationalism and expertise in the modern world. Their opponents come from a variety of mutually hostile camps. For fundamentalist religious thinkers Darwinism still represents one of the most powerful factors encouraging us to turn our backs on God and His Word. In America, fundamentalists promote a massive campaign to discredit the theory of evolution and its materialistic implications. The resulting debates reveal that even today most ordinary people still do not understand the basic principles upon which Darwin's theory is based. At a slightly less strident level, criticism of Darwinism also comes from anti-materialist thinkers who worry not so much about Genesis, but about modern biology's tendency to reduce human beings to mere puppets in the hands of natural forces. They fear that we shall lose respect for our creativity and for our status as moral beings. Their solution is to replace natural selection not with divine creation but with some more purposeful evolutionary process.[3]

Darwinism is also perceived as having political implications. Those to the left of the political spectrum still use the term 'social Darwinism' as a label for any effort to claim that human behaviour is determined by our biological nature. When conservatives argue that the existing state of society is 'natural' because it reflects our biological character, socialists invoke the memory of Darwin and the 'struggle for existence' to show that even science itself can be shaped by its political environment. Darwin, they claim, merely projected the capitalist model of a competitive society onto nature – and generations of conservatives have seized upon the model he invented to claim that their values are truly natural. In fact the conservative ethic has appealed to other biological theories besides Darwin's for justification, although it is certainly true that some modern controversies centre upon new ways of applying Darwin's insights to mankind. The science of sociobiology provoked an outcry both from social scientists and from political radicals when it was used to explain human behaviour in terms of instincts imposed on us by natural selection. The new social Darwinists, it is feared, will block reform by claiming that human nature cannot be modified by education.[4]

It would be pointless to present an analysis of Darwin's life and work without making some effort to show how the widely differing images of his theory outlined above tend to colour our interpretation of the man himself. One cannot write a value-free account of a person whose name has come to be used as an ideological symbol. Even the claim that the biographer hopes to untangle the man from the myths surrounding him must be suspect. How can the reader be sure that the writer has not built his own myth so fundamentally into the account that it will appear to the uninitiated as fact? In these circumstances the only hope is to begin, not as in a conventional biography with the subject's birth and childhood, but with a comprehensive exposition of the problems that must be faced by anyone trying to come to grips with a name that can never be freed entirely from its symbolic overtones. If the readers are made aware of what those overtones are, they have at least been warned of the need to make up their own minds on topics that are open to more than one interpretation.

What follows is not a biography in the conventional sense, although obviously I shall try to present an outline of what Darwin did and said.

Instead I shall offer the reader an opportunity to rethink his or her position on what the various facets of Darwin's life ought to mean to us today. I want to challenge the myths that have built up around Darwin by showing how historians have been forced to re-evaluate the significance of even the most basic features of his work. The result may not be a coherent and entertaining account of Darwin's life, but it will represent an open-handed effort to alert everyone to the perils inherent in taking the existing preconceptions about him for granted. To those who object that my technique is that of the historian not the biographer, I can only reply that as a historian I do not believe that a conventional biography of Darwin is necessary or even possible. The world does not need another allegedly objective account of Darwin's life, but it does need seriously to reassess his place in the growth of modern science and modern thought.

Nevertheless, it must be conceded that the time is ripe for a comprehensive reassessment of Darwin's achievement. The Darwin industry has uncovered – and is busily publishing – a vast collection of his notebooks and correspondence. The sheer extent of the material seems to have discouraged those involved from attempting a comprehensive and unified account of his life. The last serious effort at a Darwin biography by someone who really knew the material was written by Sir Gavin De Beer in 1963.[5] De Beer was a biologist who pioneered the discovery and publication of Darwin's papers, and most historians respect his efforts as a biographer even though they are aware of his limitations. Since then there have certainly been attempts by professional biographers to write a life of Darwin but these have largely failed to grapple with the problems of reinterpretation being posed by historians.[6] The time is approaching when those who are deeply involved in the Darwin papers should be willing to step back from their detailed researches and present an overview.[7] In the meantime, however, it may be useful if someone with a more detached view of Darwin's role in the development of evolutionism presents the case for a reassessment of his work.

Even in the case of less controversial figures, the biography of a creative thinker presents daunting problems. We need to look at as much evidence as has survived in order to create a plausible account of how the person gained the insights for which he or she is remembered.

Psychologically, the biographer must 'get under the skin' in an attempt to reconstruct the thought processes which made it all possible. Inevitably, this entails a careful assessment of how the subject interacted with his or her social and cultural environment. What incidents sparked off the train of creative thought, and how was the new idea presented to the world at large? In the case of Darwin, the private papers now available help to fill in many of the gaps that are missing in his own autobiography and in contemporary accounts of his work. We can see precisely how his thoughts flitted backwards and forwards as he tried to assess the various strands of evidence that he thought would bear on the question of the origin of species. We can see how his family life helped to support him and how his wife to some extent served as a sounding board by which he judged the public acceptability of his ideas. We can see how he built up a network of scientific correspondents which supplied him with information and which, in the end, became the spearhead by which his theory assaulted the public mind.

The fact that Darwin was a scientist whose theory had implications outside the realm of biology means that an assessment of his achievement poses special problems. In the popular imagination, scientists are people who make an objective study of nature and come up with factual information that forms a permanent contribution to human knowledge. Unlike the artist, the scientist produces something whose value is unquestionable – it cannot for instance rise and fall in popularity according to changing cultural fashions. Some would say that the scientist is not creative in the same sense as the artist: he merely looks carefully at nature and records accurately what he sees, imposing no interpretation of his own. A moment's reflection shows that this rigid distinction is unworkable. Although scientific hypotheses must be tested by observation and experiment, it is obvious that all the great scientific theories arose from major leaps of the imagination, from new ideas about how nature *might* work, which were only subsequently shown to have some factual validity. The scientist does not merely record factual information; he has to think up theoretical schemes that will make sense of what he sees and suggest further avenues of research. The availability of the Darwin papers allows psychologists to treat his discovery of natural selection as an important test-case for our understanding of creative thinking.[8]

However, there is a very controversial issue lying at the heart of Darwin's creation of the selection theory. Some critics have argued that the greatest fault of De Beer's biography is that it takes for granted the orthodox view of the scientist as someone who gains his inspiration solely from the factual studies in which he is engaged. For De Beer, Darwin was essentially a biologist, and the story of his discovery is the story of those lines of biological investigation which led him to see how the struggle for existence could influence genetic variability by natural selection. Darwin's non-scientific life provided merely the necessary support for his work – it could not influence his thoughts in any way. Once published, the theory of evolution by natural selection obviously affected Victorian culture because it undermined traditional religion. But the relationship between science and religion is a one-way affair: scientific facts can have a cultural spin-off, but culture cannot determine the direction of scientific discovery.

Those who criticize this image of science as a purely objective study of nature use Darwin as a classic example of how science *can* be influenced by external factors. They argue that it is too much of a coincidence that a theory portraying nature as a scene of constant struggle was created in the heyday of Victorian *laissez-faire* capitalism. Darwin's admission that he gained a crucial insight from Thomas Malthus' principle of population-expansion shows that he was fully in touch with the social philosophy of the time. He projected the competitive ethos of capitalism onto nature and then bent all his observations to fit into the pattern imposed by his own mind. Darwin did not *discover* natural selection: he *invented* it and then sold it to a world that was only too willing to see its own values provided with a 'natural' justification. The scientists' efforts to portray Darwin as a purely objective researcher are merely a device used to conceal the ideological foundations of science itself. When carried to their ultimate extremes, the approaches of both the scientists and their opponents threaten Darwin's status as a creative thinker: one because it sees him as a mere recorder of facts; the other because it reduces his ideas to nothing more than a reflection of the cultural environment within which he was immersed.

The claim that Darwin was not a creative thinker is also made by the few modern biologists who think that his theory of natural selection is simply wrong.[9] They do not want to concede that *all* science is influenced by ideology but they must invoke the social factor to explain why *bad* science can sometimes gain a hold on the public imagination. Presumably they view their own alternatives to Darwinism as truly objective, although history shows that anti-Darwinian ideas have always been associated with their own philosophical and ideological baggage. To understand someone like Darwin we must be prepared to transcend all these simple black and white alternatives. The great advantage provided by the Darwin papers is that we can see the true complexity of his thought processes *and* of the social processes by which he sought to present his theory to the world.

Modern historians see science as a process by which 'knowledge' is created rather than discovered. Obviously there is a factual input, and Darwin can be seen grappling with the many problems that he had to solve in order to reconcile his ideas with his own and other biologists' observations. At the same time it is clear that he was aware of the disturbing consequences that his ideas would have when measured by contemporary values. He thought deeply about these consequences and was fully conversant with the various approaches that others had adopted towards the wider issues. There can be little doubt that social theories played a role alongside the facts of nature in determining the direction both of his thoughts and of his search for a means by which those thoughts could be safely articulated. His contacts with other scientists can be seen as a social process by which he negotiated a format that would allow his ideas to gain wider acceptance. Darwin modified the presentation of his ideas and sought to modify other people's views on what was both scientifically and philosophically acceptable. In thus recognizing that the processes of discovery and presentation have a social dimension we are accepting that science is a human activity. We are not asserting that the scientist has unlimited freedom to create ideas that bear no relationship to reality, but we must accept that to some extent our perception of reality is a social process and that scientists cannot escape this fact.

But what, in the end, was the impact of Darwin's theory upon Victorian science and thought – and what is its legacy to the modern world? When we move on to consider this wider question we are forced once again to grapple with issues whose resolution depends on our attitudes towards the theory itself. De Beer's biography was written on the assumption that the theory of natural selection was a major breakthrough in biology. De Beer himself had participated in the creation of the 'Modern Synthesis' of Darwinism and genetics which had enabled the selection theory to dominate scientists' thinking on evolution through the 1940s and 1950s. He was writing before the start of the controversies over the adequacy of the Modern Synthesis which have led some biologists to question the adequacy of the selection theory. Not surprisingly, he presented Darwin as a good scientist who made a discovery of lasting value. To the extent that De Beer dealt with the impact of the theory, he tended to assume that the publication of the *Origin of Species* marked a turning point in the emergence of modern evolutionism. Darwin's theory laid the foundations for current thinking on the subject, although later biologists had, of course, found it necessary to make some improvements, especially to Darwin's rather primitive ideas on heredity.

New Perspectives on the Rise of Evolutionism

De Beer's account of Darwin's life is consistent with an interpretation of the history of evolutionism in which the line leading from Darwin's discovery to modern Darwinism is clearly the 'main line' of development. He was not the only scientist to adopt this approach: Loren Eiseley's *Darwin's Century* of 1958 spelled out the whole sequence through into the mid-twentieth century. While conceding that there were problems with Darwin's original formulation of his theory, particularly in the area of heredity, Eiseley nevertheless presented the subsequent history of evolutionism as the filling in of a few missing pieces in the jigsaw puzzle that Darwin had largely completed. More recently Ernst Mayr, another founder of the Modern Synthesis, has provided an authoritative statement of the same interpretation.[10] Mayr concedes that other ideas about evolution were proposed during the nineteenth century but he devotes little space to them. They are thus dismissed as side branches leading away from the main line, which is firmly Darwinian in character. To a large extent, the incredibly detailed work of the modern Darwin industry is a reflection of this view of history in which Darwin plays the central role in the emergence of modern evolutionism.

With this model of the history of evolutionism, if we turn to look at the theory's wider impact it is only natural to assume that the selection theory played a major role in shaping the transition from creationism to social Darwinism in the late nineteenth century. Even those historians who do not like the implications of Darwinism seem to accept this view. Biologists such as Lovtrup dismiss scientific Darwinism as a confidence trick imposed on the profession by ideological pressure in the age of capitalism. Cultural historians such as Jacques Barzun and Gertrude Himmelfarb blame the *Origin of Species* for ushering in an age of materialism and the worship of brute force, thus paving the way for the horrors of the twentieth century.[11] Such an interpretation still reflects the metaphor of a war between materialistic science and the spiritual values represented by traditional religion. The claim that the 'Darwinian Revolution' in science helped to precipitate a major cultural transition also seems to reinforce the view that Darwin merely reflected the ideology of his time. It throws the responsibility for the origins of much modern nastiness back onto the shoulders of the man who popularized the metaphors of the 'struggle for existence' and the 'survival of the fittest'. It hardly seems to matter whether you love Darwin's message or hate it; you cannot escape the fact that it helped to overturn the traditional Christian world view.

Unfortunately for the advocates of this view, modern historical research is beginning to show that the impact of Darwin's theory on science, let alone on Western culture as a whole, was a good deal more complex than we used to imagine. There were many varieties of nineteenth-century evolutionism, none of which corresponded unequivocally to modern Darwinism. The metaphor of a war between science and religion, with evolution invariably coming in on the materialists' side, has proved to be seriously misleading. Drawing upon this research I have argued that the image of a Darwinian Revolution is a myth created by modern Darwinists, a myth which has gone largely unchallenged because even their opponents find it useful as a means of exaggerating the theory's dangers.[12] The simple fact is that Darwin's theory of natural selection remained highly controversial throughout the late nineteenth century. Only a handful of biologists took Darwin's mechanism seriously, while the vast majority opted for various anti--Darwinian ideas and dismissed natural selection as a secondary, purely negative factor. Selection explained the elimination of the unfit but not the origin of the fitter organisms which replaced them. Even the great social Darwinists either ignored or misrepresented what we now regard as the most important aspects of Darwin's thought. Far from constituting the main line of development in nineteenth-century evolutionism, Darwin's theory was in some respects an anomaly that his contemporaries were unable to come to grips with. In a sense Darwin was 'ahead of his time', because his more radical proposals were only taken seriously after biology had been revolutionized by the emergence of Mendelian genetics.

Given the widespread popular misunderstanding of Darwinism, it is important for us to be clear about the revolutionary nature of Darwin's proposals as they are understood by modern biologists. The essence of natural selection is Darwin's suggestion that evolution is guided solely by the interaction between the population and its environment. The members of a population differ among themselves in a host of minor ways which animals (and humans) use to identify one another as individuals. This more or less random variation (which we now explain in terms of genetic differences) forms the raw material upon which natural selection works. Selection involves the preferential survival and reproduction of those individuals which by chance have inherited a variation that gives them an edge over their neighbours in coping with the local environment. These 'fitter' (that is, better adapted) individuals survive and breed more readily than the others and their advantageous character thus increases in frequency in the next generation. Over a long period the adaptive feature spreads through the whole population and the average character of the species changes. Darwin and the modern Darwinians assume that small changes produced in this way will eventually add up to give the major developments that have characterized the appearance of new forms of life in the course of the earth's history.

The crucial assumption is that the individual variation upon which natural selection operates is essentially random: by itself it cannot force evolution along a particular path because its tendency is to spread out in all directions. In Darwin's theory there can be no inbuilt trend forcing the species to evolve in a certain direction. In particular there is no force that compels species to progress along a preordained hierarchy of complexity, no evolutionary ladder that all species must ascend. Because it is directed solely by the demands of the local environment upon the population, evolution is an essentially open-ended process with no single goal. If a sample of an original population is transported to an isolated location such as an oceanic island it will adapt to its new environment as best it can. As Darwin observed in the Galapagos islands, when samples are transported to several different but isolated locations, each will change without reference to the others and the result will be a group of different but related 'daughter' species. Further evolution will produce yet more branching and divergence. With such a model of evolution, it is impossible to pick out one modern species such as the human race and say that this represents the goal towards which the whole process has been working. Each branch is different, and the biologist cannot evaluate the products of one branch using criteria derived from another.

It is precisely this open-ended and non-progressive aspect of Darwin's theory that his opponents have always found unacceptable. In the later nineteenth century many 'Darwinists' paid lip service to the idea of natural selection, but they preferred to believe that evolution was, after all, guided inexorably in the direction of progress towards mankind. Opponents of Darwinism openly sought alternatives that would explain why variation was directed along a preordained path as though towards a purposeful goal. Even when the idea of progress became less attractive, biologists still maintained that evolution must somehow be directed along fixed paths by inbuilt biological trends. The more extreme modern opponents of Darwinism such as Lovtrup still argue for some form of internal, that is, non-environmental, control over the direction of evolution. There have been many biologists and social thinkers who have objected to the role played by the 'struggle for existence' in Darwin's theory. But in fact this is rather a red herring. There are other theories besides Darwin's which have included a role for struggle (and which are often mistakenly called 'Darwinian'). The real source of the opposition to the selection theory has been a deep-rooted objection by many thinkers to the idea that evolution could be an open-ended trial and error process. The opponents want evolution to be directed towards a future goal, although they cannot agree among themselves whether it is to be directed by God, by the striving of individual animals or by inbuilt genetic factors.

It is in the context of this distinction that it becomes possible to argue that what we now perceive to be the core of Darwin's theory was not taken up by his contemporaries. The *Origin of Species* converted the world to evolutionism but it did so despite Darwin's failure to convince his contemporaries that natural selection was an adequate mechanism to explain the process. Only after the modern concept of genetic mutation confirmed that individual variation is essentially random were biologists forced to come to grips with the prospect that evolution might have the open-ended character predicted by Darwin. If this is so, then the question of how Darwin interacted with his contemporaries becomes absolutely crucial. In the interpretation offered by De Beer and the modern Darwinians, it is simply assumed that Darwin converted the world to evolutionism because the arguments contained in the *Origin of Species* were so convincing. In the interpretation of their opponents, he succeeded because he merely reflected the values of his time. But if the *Origin of Species* had an important role to play despite the failure of Darwin's chief explanatory tool, we need to ask what that role was and how Darwin was able to play it. The claim that scientific 'discovery' is a social process becomes all the more important, and Darwin's skill as an advocate becomes just as crucial as his basic ideas.

The thesis of my new interpretation of Darwin's role is that the *Origin of Species* acted as a catalyst which precipitated the conversion of many nineteenth-century thinkers to a progressionist version of evolutionism that was not 'Darwinian' in the modern sense of that term. The term 'Darwinism' has to be understood in two entirely different ways, one appropriate for the Victorian era and one for the twentieth century. To a large extent historians' fascination with the discovery of natural selection is a product of the latter definition: we are interested because we know with hindsight that the theory would eventually prove to be of immense influence within modern biology.

That is surely a legitimate interest, but it must not be allowed to blind us to the necessity of evaluating the impact that Darwin had on his own time.

Once we realize that the selection theory did not carry all before it when first published, we need to ask a series of entirely new questions about the way in which Darwin interacted with his contemporaries. To what extent had nineteenth-century science and culture begun to move towards an evolutionary viewpoint before the *Origin of Species* appeared? Why had that movement become blocked so that a book containing an unacceptably radical idea was able to act as the catalyst that precipitated the transition? To what extent did Darwin's skill at presenting his theory to a scientific and a general audience rest upon his willingness to adapt his language and metaphors to the prevailing values of the time? Why did Darwin's name become a symbol for progressionist evolutionism if modern biologists regard his theory as undermining the whole logic of progressionism? To answer these questions we need to get away from the old stereotypes in which Darwin succeeded either because he got the right answer or because he got the wrong answer that everyone preferred to hear about. We need to reassess why we think he is important both for modern science and for modern thought. Just as modern scientific Darwinism goes beyond anything that was possible in the nineteenth century, so it must become obvious that the original form of Darwinism did not immediately usher in an age of rampant atheism.

A meaningful life of Darwin must try to answer the questions outlined above. We cannot describe the discovery and publication of his theory in a vacuum as though its value were self-evident. Darwin's work can only be understood in the context of his time: how he absorbed, transmuted and in some cases transcended contemporary values to create a theory that would only be fully appreciated long after his death, yet would be exploited as a symbol of the progressionist ideology of his own time. We need to interweave a study of how he was led to his most radical insights with an appreciation of the extent to which he responded to the demands of his scientific and cultural environment.

Darwin is an unusual figure in the history of science because he was hailed as a hero in his own lifetime, yet not for those achievements for which he is most respected today. He must be evaluated throughout from two different and potentially contradictory perspectives. We must constantly balance the need to understand the discoveries respected by the modern biologist against the historian's desire to uncover the complex role played by 'Darwinism' in the development of nineteenth-century thought. Darwin's life has a dual meaning for us today: he pioneered important scientific insights, but he also affected the cultural development of his own age in a way that influenced, but did not simply anticipate, the emergence of twentieth-century values. For this reason, we begin our study not with Darwin's birth and childhood but with a brief survey of how the problem of evolution was viewed by the rest of his society during the time in which he was secretly developing his ideas.

2

Evolution before the Origin of Species

I N 1831 D ARWIN left England aboard the survey ship
HMS *Beagle* at the start of the round-the-world voyage
that would provide him with his first evidence for evolution. He
returned five years later and soon began a series of investigations
designed to reveal whether a natural mechanism of transmutation was
conceivable. The basic idea of natural selection first occurred to him in
1838. Over the next two decades Darwin worked on his theory in
secret, telling only a handful of close friends what he had in mind. In
the 1850s he began to write out a long multi-volumed account of his
theory, but this process was interrupted in 1858 by the arrival of a
paper by Alfred Russel Wallace outlining a theory very similar to his
own. Darwin and his friends arranged for extracts of his own theory
to be published along with Wallace's paper by the Linnean Society.
Almost immediately Darwin began writing the account of his theory
that was published at the end of 1859 under the title *On the Origin of
Species*.

Traditional accounts of Darwin's discovery tend to imply that very little attention was being paid to the idea of evolution by other biologists during the decades leading up to 1859. It is assumed that almost everyone accepted a fairly straightforward creationism and that the vast majority of biologists went out of their way to argue that the adaptation of each species to its environment proved the existence of a wise and benevolent Creator. William Paley's *Natural Theology* of 1802 is seen as a classic exposition of this 'argument from design' – the claim that species are designed by an intelligent Creator in the way that a watch is designed and built by a watchmaker. Naturalists accepted this position because the alternative of natural evolution was inconceivable, and even Darwin had his attention focused on the problem of adaptation by his early interest in Paley's book.

Opponents of Darwinism have always complained about the one-sidedness of the assumption that Darwin's was the first real theory of evolution. In 1879 the novelist Samuel Butler, who had conceived a violent distaste for the theory of natural selection, published his *Evolution, Old and New* to argue that Darwin was by no means the first to advance a theory of evolution. Butler pointed out that a number of earlier naturalists had openly suggested natural processes of organic change. These included the French biologists Georges Buffon and J. B. Lamarck and Darwin's own grandfather, Erasmus Darwin, whose *Zoonomia* of 1794–6 had contained a chapter on transmutation. In 1844 the anonymously published *Vestiges of the Natural History of Creation* (actually by the Edinburgh writer Robert Chambers) had generated a public outcry with its suggestion that mankind had emerged from the lower animals.

Writers more sympathetic to Darwin have always conceded that some earlier statements on evolutionism were made, but argue that no pre-Darwinian naturalist was able to generate widespread support for the idea. Lamarck, who proposed a comprehensive theory of natural development in his *Zoological Philosophy* of 1809, was ridiculed by the great anatomist Georges Cuvier and extensively criticized in the second volume of Charles Lyell's *Principles of Geology* (1830–3). The fact that Lyell was a great influence on Darwin shows that even those naturalists with potentially the most to gain from the theory were unwilling to take it seriously. Chambers' book did more harm than good because it offended people by linking mankind to the animals but offered no plausible mechanism of evolution. Thus, it is argued, Darwin's *Origin of Species* really did appear as a breakthrough.

Modern historical research has shown that both these positions are caricatures which obscure the complex role played by the idea of evolution in the early decades of the nineteenth century.[1] It is worth asking why, if Lamarck's views were totally ignored, Lyell and a number of other scientists made such a fuss about refuting him. It is now clear that a number of important developments were taking place alongside Darwin's secret researches. Some of these developments were not directly related to the idea of evolution, but helped to pave the way for the developmental view of nature that Darwin would have to interact with. In the course of the half-century from 1800 to 1850, the modern outline of the fossil record had been put together, revealing the ascent of life from primitive fish and invertebrates through the Age of Reptiles to the Age of Mammals and the world of today. Even creationism had to become much more complex and sophisticated if it was to cope with this flood of information on the development of life.

In addition, we now know that Lamarck's theories were not completely ignored; there was a radical movement both within and outside the realm of science that was prepared to exploit such materialistic ideas as part of a campaign against the establishment. It is clear that Darwin went out of his way to dissociate himself from this radical movement. Those who attacked evolutionism had a definite target in mind, and to some extent the more sophisticated versions of creationism were a response to the radical threat. The debate over Chambers' *Vestiges* shows that by the mid-1840s efforts were underway to create a less ideologically suspect form of evolutionism. Darwin's theory was thus injected into a culture that was fully aware of what transmutation implied – and Darwin himself was fully aware of the need to accommodate his theory to the preferences of the time. Trying to understand the development and presentation of Darwin's theory without reference to these earlier debates can only lead to misunderstanding and over-simplification.

Radical Evolutionism

The work of Adrian Desmond has revealed how thoroughly politicized were the biological debates of the early nineteenth century.[2] Outside the ranks of both science and polite society were political radicals and revolutionaries inspired by the events in France and determined to bring down the aristocracy in Britain. Their illegal pamphlets and broadsheets often used materialist ideas to support their campaign against the Church, which was seen as a bastion of the State. They appealed to transmutationism to support the claim that humans are nothing more than highly developed animals, thus undermining the belief that we have a soul capable of appreciating divinely revealed truths, a belief which always seemed to imply that one should support the existing social hierarchy. The writers of the gutter press may not have understood the details of Lamarck's arguments, but they knew that his theory had implications that were directly relevant to their case.

Lamarck's theory was based on a number of assumptions that Darwin would reject, at least partly as a means of distancing himself from the materialist label that was attached to them. To begin with, Lamarck actively linked his ideas on the development of life to the claim that the first living things had been produced by 'spontaneous generation', that is, by a natural transition from non-living to living matter.[3] To conservative thinkers this claim seemed to strike at the heart of the traditional belief that life was the gift of a divine Creator. In addition, Lamarck held that the simplest forms of life advanced gradually but inevitably along a scale or hierarchy of complexity, until at length the human race was formed. The idea of necessary progress was thus widely assumed to imply that mankind was merely a highly developed animal. Both spontaneous generation and the idea of a linear progressive trend became integral parts of the materialistic evolutionism that so concerned the scientific world of early Victorian Britain.

For later biologists, however, Lamarck's name has become associated more closely with what was actually only a minor part of his theory, the mechanism of the 'inheritance of acquired characteristics'. This hypothetical mechanism of adaptive evolution is based on the assumption that the effects of bodily changes in the adult animal can be passed on to the offspring and can thus accumulate to transform the species. To use one of Lamarck's best-known examples, the efforts made by generations of giraffes trying to reach the leaves of trees are supposed to have gradually lengthened their necks until the modern species was formed. Darwin himself continued to give some credence to this mechanism, although he subordinated it to natural selection. In the later nineteenth century some of Darwin's opponents began to call themselves 'Lamarckians', but by this time the inheritance of acquired characters had been dissociated from the rest of Lamarck's views and from the materialist label attached to them. Modern genetics suggests that, in fact, characters acquired by an adult's efforts cannot be transmitted to its children.

Because the idea of transmutation was exploited by some early nineteenth-century revolutionaries, it was difficult for less extreme radicals to use the theory without being tarred with the same brush. Nevertheless, there were some radically minded biologists who were prepared to take the risk. Writing of his days as a medical student in Edinburgh, Darwin later recalled his acquaintance with the young lecturer Robert Edmond Grant, who 'one day, when we were walking together burst forth in high admiration of Lamarck and his views on evolution'.[4] In his *Autobiography* Darwin hastened to distance himself from Grant by continuing: 'I listened in silent astonishment, and as far as I can judge, without any effect on my mind'. As a result, historians have tended to dismiss Grant as a minor figure, an unlucky precursor of evolutionism who could not gain a hearing. More recent research has now begun to suggest a very different picture. At the beginning of his career Grant was seen as a future leader in his field and his radical opinions were widely circulated. He may not have converted the young Darwin to evolutionism in 1826, but he certainly influenced Darwin's thinking by suggesting important topics of research.[5]

Grant had been trained in medicine at Edinburgh and studied for a time in Paris after the fall of Napoleon. Here he may well have met Lamarck and certainly became acquainted with his teachings. He continued to visit Paris regularly, and in 1823 was appointed to teach an extramural course on the invertebrates at the University of Edinburgh. There is evidence that he made no secret of his transformist views when teaching or in discussions on natural history at the Plinian Society. An anonymous article supporting Lamarck in the *Edinburgh New Philosophical Journal* for 1826 is almost certainly by Grant.[6] Adrian Desmond suggests that radically materialist views on both scientific and political matters must have circulated more freely in the Edinburgh of the 1820s than we normally suspect: only south of the border were such ideas considered totally beyond the pale.

In 1827 Grant was appointed Professor of Anatomy and Zoology at the newly founded University of London. His appointment was a sign of his growing scientific expertise, but here he had to be far more circumspect in the expression of his radical views. Nevertheless, his lectures made his position clear to anyone who was prepared to read between the lines. He was also able to link up with groups determined to reform the medical profession. The medical establishment, including the Royal College of Surgeons, was committed to the old teleological view of nature in which every species (including, of course, the human race) was designed by a wise and benevolent God. Moves were now afoot to broaden access to medical education through the creation of private medical schools. An uneasy alliance of religious nonconformists and political radicals set out to challenge the establishment's monopoly. They favoured the new more radical approach to natural history because it allowed them to dismiss establishment science as outdated and out of touch with the latest continental ideas. The medical reformers were quite willing to listen to Grant's attacks on teleology and his emphasis on the linear pattern of development revealed by the history of life.

Grant thus came into conflict with established medical interests, and throughout the 1830s he was engaged in scientific debates which had a hidden agenda. Conservative biologists began to update their image and look for new ways of discrediting the Lamarckism which threatened their traditional beliefs. Grant was increasingly branded as an outsider whose views were unacceptable in polite society. Philosophical materialism implied ethical materialism and hence the destruction of the moral order upon which society was built. In the politically unstable years of the 1830s and 1840s, it was all too easy to say that a materialist must be a social revolutionary. Grant's position at the University was also a difficult one: lack of money forced him to undertake vast amounts of teaching, and his research began to suffer. During the 1840s he was effectively marginalized, both professionally and socially. He ended up living in a slum, a discredited figure who served as a clear warning to those (such as Darwin) whose scientific views threatened the intellectual pillars upholding the social order.

It was during the mid-1840s that the debate over the implications of evolutionism began to gain wider attention. The detonator for this explosion of interest was a book published anonymously in 1844 under the title *Vestiges of the Natural History of Creation*. The author's name was a closely guarded secret and there was much speculation over whom the culprit might be. In fact, the book had been written by the Edinburgh author and publisher Robert Chambers, and thanks to the work of James Secord we now know a great deal about the events leading to the book's publication.[7] Older histories of evolutionism invariably dismiss *Vestiges* as yet another failed attempt to anticipate the logic of Darwinism – one account of the book's role is entitled *Just before Darwin*.[8] It is assumed that Chambers' lack of scientific knowledge led him to mount an inadequate campaign in support of evolution, a campaign that was doomed to failure because he made no secret of the theory's implications for the origins and status of mankind. Closer reading shows that *Vestiges* is in no sense a forerunner of the *Origin of Species*. It argues for a very different mechanism of development in which a divinely instituted progressive trend forces life to mount steadily up the scale of organization towards higher intelligence.[9] It is now possible to see *Vestiges* as an attempt to introduce issues that were already widely discussed in Edinburgh to an audience living in the more conservative atmosphere south of the border. Far from being a failed anticipation of Darwinism, *Vestiges* established the agenda for all future discussion of evolutionism, including the *Origin of Species*.

Chambers' purpose was to dissociate evolutionism from its radical image and make it acceptable to the growing middle class. For some years his *Chambers' Edinburgh Journal* had been promoting the message that social progress was inevitable if only people could be given the freedom to innovate. The key to success should be effort and initiative, not aristocratic privilege, and if individuals were given the freedom to exert themselves, society itself would reap the economic and technological benefits. This was the natural social philosophy of the new commercial and industrial entrepreneurs, and the purpose of *Vestiges* was to argue that social progress is inevitable in the long run because it is a natural extension of the progressive development of life in the course of the earth's history. Precisely for this reason, Chambers made explicit the link between mankind and the animals implied by a theory of transmutation. He adopted the ideas of the Scots phrenologist George Combe, whose *Of the Constitution of Man* of 1828 had argued that the physical structure of the brain is the source of all mental functions. Phrenology was eventually discredited as a pseudo-science because it assumed that a person's character could be read from the bumps on his or her head.[10] But in its early years it was a powerful factor in the promotion of materialist values. Chambers used it to argue that an expansion of the animal brain in the course of progressive evolution would inevitably lead to the production of the human mind.[11]

Chambers attempted to soften the blow by arguing that the progressive trend could itself be seen as a continuously operating divine plan of creation. God established the law of development which led to the appearance of successively higher species, instead of creating each new arrival Himself. *Vestiges* was thus presented as a modernization of the old teleology: mankind's unique status was proclaimed not because we are a direct divine creation but because we are the goal of all natural development. Although not a Lamarckian in the sense that he accepted the inheritance of acquired characteristics, Chambers nevertheless retained the linear progressionism that Grant had associated with the theory. He argued that the growth of the human embryo reveals a speeded-up version of the development of life on earth, thus anticipating the 'recapitulation theory' that was to become popular in post-Darwinian times. However, for Chambers, the link between evolution and embryology was a fundamental clue to the progressive character of all natural development. Evolution progressed inevitably towards its goal just as the embryo grows steadily towards maturity.

There was a predictable outcry against *Vestiges* from conservative thinkers. The Professor of Geology at Cambridge, Adam Sedgwick, wrote an eighty-five-page review protesting that 'our glorious maidens and matrons' must be protected from this poisonous nonsense.[12] Yet there were signs that attitudes were changing. The anatomist Richard Owen, who had led the campaign against Grant in the 1830s, did not speak out against *Vestiges*. He wrote openly in his *On the Nature of Limbs* of 1849 that the divine plan of creation might be unfolded through secondary, that is, non-miraculous causes.[13] In this, Owen went too far for his conservative backers and he avoided the topic of the origin of new species throughout the 1850s. His willingness to think about the possibility of transmutation, coupled with his silence in print, is typical of the period leading up to the publication of the *Origin of Species*. People were now alerted to the possibility that one might abandon simple creationism for a theory in which evolution expressed a divine purpose, without thereby giving way to radical materialism. But conservative scientists were unable to explore this idea because their supporters were as yet unwilling to face up to the rethinking of traditional values that would be required.

At the more liberal end of the political spectrum Chambers' ideas seemed to offer a potential way forward. Once the initial outcry had died down, liberal Anglicans such as the Oxford mathematician and philosopher Baden Powell began to move in the direction sketched in by *Vestiges*.[14] Powell's *Essays on the Spirit of the Inductive Philosophy* of 1855 expressed the view that God's government of the universe was more apparent in the laws He had instituted than in arbitrary miraculous violations of those laws. Powell was openly linking this view to the question of the development of life on earth, and we can see his position as an anticipation of one of the most common reactions to the *Origin of Species*: acceptance of evolution provided that the mechanism of change retains the element of teleology denied by Darwin. In a sense, the new initiative offered in Darwin's book pre-empted a gradual move towards teleological evolutionism that was already under way in the decades following the publication of *Vestiges*.

At a yet more radical level, there were a few thinkers who already felt that Chambers had not gone far enough. The young social philosopher Herbert Spencer had begun sketching in the outlines of the position that would lead later commentators to brand him as a 'social Darwinist'.[15] In his *Social Statics* of 1851 Spencer emphasized the need for free enterprise to guarantee the adaptation of the individual to an ever-changing society. He extolled the sufferings that are a consequence of failure as the best possible stimulus encouraging the individual to do better next time. Contrary to popular belief, Spencer's emphasis on the progressionist consequences of *laissez-faire* individualism was not a direct anticipation of natural selection. He was more concerned with the individual's positive response to an environmental challenge, and from the first he suspected that such responses could be inherited and thus be incorporated into the species' biological character. In an essay in 1851 on 'The Development Hypothesis' Spencer proclaimed his support for Lamarck's mechanism of the inheritance of acquired characteristics.[16] Like Chambers, he linked biological and social progress into a single developmental scheme, but his Lamarckism allowed him to argue that the mechanism of change at both levels is not a mysterious divine plan but the cumulative efforts of individuals responding to their environment.

As yet, few scientists were willing to take another look at Lamarckism, even though Spencer showed how it fitted neatly into the ideological values of the commercial entrepreneur. Most professional naturalists felt that the theory had been too firmly discredited in the previous generation, and they still feared the label of materialism. The stalemate forced on science by the collapse of Lamarckism can be seen most clearly in the attitude of a man who would later become a champion of Darwinism: Thomas Henry Huxley. In the mid-1850s Huxley was a clever newcomer desperately trying to find a niche for himself as a professional scientist.[17] He saw the hollowness of creationism and hoped that science would soon be able to penetrate into this area which had so long been claimed by theologians. But he could not take Lamarckism seriously, while he dismissed *Vestiges* as a scientifically meaningless twist on the old idea of a divine plan of creation.[18] As he later wrote of his own reaction to the *Origin of Species*, he had been inclined to say to creationists and transmutationists 'a plague on both your houses'.[19] The *Origin of Species* seemed a breakthrough not because Huxley was unaware of the potential value of evolutionism, but because he was blocked from considering the idea unless he saw that a new approach to the problem of *how* transmutation worked was possible.

We can thus see the delicate state of affairs into which Darwin eventually injected his theory. Even some conservative scientists had become aware of the need to postulate some sort of developmental process in nature, but they were forced to tread carefully because any suggestion that mankind was merely a highly developed ape raised theological problems that few were as yet willing to face. More radical thinkers were anxious to challenge the Church's monopoly on explaining the origin of life and wanted a popular philosophy of progress in which future reform could be presented as the inevitable outcome of a universal trend. But here too the way forward was blocked, because the blacklisting of Lamarckism meant that there was no respectable mechanism of evolution upon which to build the revolution in science that would be the first step toward imposing a progressionist world view upon Victorian culture. In these circumstances, a new initiative by a respected naturalist was bound to have major repercussions.

The Opponents of Transmutation

The gradual development of radical evolutionism in the pre-Darwinian decades forced even those who opposed the idea to update their thinking in ways that would profoundly affect both Darwin's own thinking and the eventual reception of his theory. The simple-minded creationism underlying Paley's *Natural Theology* became so obviously outdated that anyone wishing to argue seriously against the new ideas was forced to think of new ways by which the traditional link between divine providence and the living world could be preserved. By the time Darwin published, even the creationists conceded that the appearance of new species followed an intelligible pattern, and in some cases the patterns they saw were remarkably close to those that an evolutionist would explain in terms of descent with modification.

Paley himself may have been reacting against earlier evolutionary suggestions put forward by Erasmus Darwin and others. He argued that no natural process could explain the creation of complex organic structures. Hence these structures in all their bewildering variety must each be the result of an act of divine creation. But already, by the time Paley published his book, it was becoming clear that the traditional belief in a single period of creative activity 'in the beginning' was untenable. The French biologist Georges Cuvier made careful studies of many fossilized creatures and showed that they were unlike any species now alive. Cuvier forced his contemporaries to accept that some species had become extinct in the course of the earth's history. He and his followers also showed that there must have been a whole series of distinct populations replacing one another in the course of geological time. If the concept of divine creation was to be retained, it would have to be accepted that the Creator had acted over and over again to repopulate the earth after its original inhabitants had become extinct.[20]

The most obvious way out of the problem was to assume that geological catastrophes had periodically wiped out all or most species, leaving the earth free for the Almighty to step in and create a series of new species adapted to the prevailing conditions. One of the leading exponents of this view was the Reader in Geology at Oxford, William Buckland, whose popular lectures helped to generate wide interest in the fossil record. Buckland has all too often been dismissed as a backward thinker who practised bad science in order to satisfy his religious preconceptions. In fact he was extremely active in the movement that helped to create the outline of the earth's history that we still accept today. Catastrophism was not necessarily a 'bad' theory designed to do little more than uphold a belief in divine creation and the reality of Noah's flood. On the contrary, catastrophists were active in creating the modern system of geological periods and in exploring the sequence of new species that appeared in the course of life's development. In particular, they outlined the progressive character of the fossil record, which seemed to begin with primitive invertebrates and then move on successively to the ages dominated by fish, reptiles and finally mammals. Buckland's *Bridgewater Treatise* of 1837 was a comprehensive survey of the fossil record, each species being described as an example of divine workmanship, perfectly adapted to the lifestyle it employed to gain a livelihood in the prevailing environment of its time.[21]

Despite Buckland's skill at interpreting individual species, it was becoming increasingly obvious that his piecemeal approach was unlikely to uncover any underlying trends that might govern the sequence of creations. The conservative forces that wished to defend Paley's view of divine providence realized that a new initiative was needed if their whole approach was not to be dismissed as outdated in the face of the latest continental developments. This was the age of the 'philosophical naturalists' who hoped to go beyond the mere description of species to uncover the underlying logic of the Creator's plan.[22] It was known that Lamarck's younger colleague Geoffroy Saint Hilaire had developed an alternative approach which stressed the fundamental unity of all animal species. In Geoffroy's 'transcendental anatomy', each species was seen as a different variant upon a basic archetypal pattern; interest was focused on the underlying unity of type, not on the variety of adaptations. In France, this approach was seen as having radical implications, not least because Geoffroy argued that new variants on the archetypal pattern might be produced by natural means. In effect, he proposed a theory of evolution by sudden saltations or 'leaps' caused when changed conditions disturbed the process of growth and led to the formation of new organic structures.[23]

Geoffroy's transcendental anatomy was taken up by another radical Scots naturalist, Robert Knox, best known nowadays for his association with the body snatchers Burke and Hare. But the basic logic of the new approach was open to a more conservative interpretation in which the existence of an underlying archetype linking all animal forms was seen as evidence that the Creator worked to a rational plan. German naturalists were already exploring transcendental anatomy in a more teleological context, and one man in particular became associated with the development of this interpretation in Britain. The promising young anatomist Richard Owen was appointed to a curatorship and (in 1836) to a professorial chair at the Royal College of Surgeons. From this bastion of medical orthodoxy he was expected to lead the assault on the radicals by showing that new ideas, when properly understood, could be incorporated into the teleological view of nature. In the course of the 1830s Owen clashed openly with Grant, arguing that neither comparative anatomy nor palaeontology was compatible with the idea of a linear progressive trend leading towards mankind. In the second part of his study of British fossil reptiles (1840–1), he coined the term 'dinosaur' to denote the great terrestrial reptiles of the Mesozoic era. Adrian Desmond has shown how Owen's concept of the dinosaur was part of his campaign against Grant: the dinosaurs were the most advanced reptiles, yet they were among the earliest members of their class. They thus revealed a trend towards degeneration rather than progress in the fossil record.[24]

The claim that the fossil record revealed progress only in a series of discontinuous steps was one of the commonest arguments used against transmutation in the 1840s. It featured prominently in Sedgwick's massive and hostile review of Chambers' *Vestiges* and in books by the Scots stonemason-turned-geologist Hugh Miller.[25] We know that Darwin read these attacks with concern, since they indicated the kind of arguments that conservative thinkers would bring against his own very different theory of transmutation.[26] He realized how important it would be for him to emphasize the imperfection of the geological record so that the apparently sudden leaps in the level of organization could be explained in terms of missing fossils. Darwin also realized that it was the idea of a linear progressive trend that was perceived as being particularly dangerous, because it seemed to imply that mankind was the end-product of development within the animal kingdom. Since his own theory did not include the idea of inevitable progress, he warned himself against using the language of simple progressionism when presenting it.

At the same time, Owen himself was becoming less committed to the discontinuity of the fossil record and by 1849 he hinted openly that the appearance of new forms might be non-miraculous. The reason for his change of heart was his exploitation of transcendental anatomy to highlight other kinds of trends in the fossil record. Owen proposed his concept of the vertebrate archetype, an idealized model of the most basic vertebrate form, which could be seen as an underlying pattern unifying all the diverse species within the type. He defined the modern concept of 'homology', noting that the same pattern of bones can be seen adapted to different purposes in the arm of a human being, the paddle of a dolphin and the wing of a bat. According to Owen, this underlying unity showed that the adaptation of structure to function was not the best indication of the Creator's wisdom: it was the rationality of the overall pattern that gave the clearest proof that nature was not just a random conglomeration of forms produced by accidental causes.

These points were made in Owen's *On the Nature of Limbs* of 1849, and it was in the conclusion of this book that he hinted at the possibility of secondary causes being responsible for the unfolding of the arche-type's various manifestations in the course of geological time.[27] As we have already noted, he felt it prudent not to repeat this claim, but during the 1850s he continued to make suggestions about the pattern of development that could be seen in the fossil record. In particular he argued that many classes exhibit a process of branching development: the earliest members of the class are usually rather generalized forms, but in the course of geological time their replacements (or descendants) branch out into numerous divergent lines of specialization, each adapting itself to a different way of life.[28] As we shall see, Darwin spent much of his effort in the 1850s trying to ensure that his theory would explain this kind of pattern, and in the *Origin of Species* he referred to Owen's views as indirect support for his own theory. Divergent lines of specialization were exactly what one would expect to result from the operations of natural selection.

Owen, of course, resisted Darwin's suggestion and attacked the selection theory as a threat to the teleological view of natural development. For him, the divergent lines represented the unfolding of the various potentials that the Creator had built into the original members of the class. Nevertheless, we must recognize the falsity of the assumption, made all too often by an earlier generation of historians, that Owen's opposition to natural selection meant that he was an anti-evolutionist. Once the barrier had been broken by the *Origin of Species* debate, Owen came out openly in support of transmutation, provided that it was seen as the unfolding of a divine plan. His position illustrates the option that was most often taken up by conservative naturalists in the late nineteenth century. Chambers' theory of a divinely preordained evolutionary pattern was dissociated from the idea of a linear progress towards mankind and used as a compromise between simple creationism and Darwin's purely naturalistic alternative. The fact that Owen himself had almost certainly been thinking along these lines even before the *Origin of Species* appeared suggests once again that Darwin's book was not introduced to a scientific community that was entirely unprepared to confront the basic idea of evolution.

The suggestion that the development of life on earth consists of something more complex than the ascent of a ladder towards mankind represents an interesting, if superficial, link between the conservative approach and Darwinism. In fact, Darwin was anxious to imply that his mechanism of natural selection would, at least in the long run, drive most living things towards higher levels of development. Only in this way could he appeal to liberal thinkers who looked to evolution as a guarantee of social progress. He also knew that most of his contemporaries saw a progressive sequence in the fossil record. But far more than the majority of his contemporaries, Darwin knew that evolution was best represented as a branching tree rather than a ladder. His progressionist followers themselves worked out a compromise in which evolution *was* seen as necessarily progressive and the image of a branching tree was modified by giving the tree a main stem or trunk leading towards mankind as the high point of creation. The other lines of development were merely side branches, interesting enough to the naturalist, no doubt, but of little overall consequence. It is only Darwin's modern followers who have been able to come to grips properly with the implications of a theory in which evolution has no preordained goal towards which it must be supposed to advance.

3

The Young Darwin

O NE THING IS now clear: Darwin did not introduce the idea of evolution to a scientific community (or a general public) that was unaware of the theory's implications. This makes any assessment of his early career doubly problematic. Even on the conventional interpretation of Darwin as a hero of discovery, it is necessary to specify those areas of investigation that led him towards so revolutionary an explanation of organic origins. But once it is recognized that he was forced to interact with other naturalists who were deeply concerned about exactly the issues that were becoming the chief focus of his attention, it becomes important to ask about the social relationships between the young Darwin and the rest of the scientific community. A realistic account of how he was led to develop the theory of evolution by natural selection must thus include both his scientific investigations and his dealings with others. The fact that Darwin refused to publish his theory for two decades after his original insight has always led to speculation about his fear of criticism. However, modern scholarship has made us far more aware of the complexities of the process by which Darwin both developed his theory and negotiated its introduction to the scientific community.

Before outlining the early stages of Darwin's career, it is thus important for us to be clear about the kind of questions we need to ask if we are to gain a full appreciation of his actions. To what extent did the radical intellectual background of the Darwin family (clearly illustrated by the opinions of Erasmus Darwin) predispose him to challenge the orthodox view of nature? How did this radical background interact with the more conventional education he received at Cambridge? How did the young Darwin gain recognition as a promising naturalist who could begin to speak on equal terms with the great names in the field? How did he set about gathering information on a topic that was likely to arouse strong opinions in almost anyone who knew what he was driving at? To what extent did his awareness of transmutationism's radical implications shape his quest for ideas and information that would be relevant to his project? Once the theory had begun to emerge, did he find it necessary to modify his views in order to accommodate either scientific or popular prejudices? These are just a few of the topics which need to be addressed in addition to the conventional story of Darwin's voyage on the *Beagle* and his investigation of animal breeding.

The historian seeking to evaluate Darwin's career is now confronted with an almost overwhelming wealth of sources. In addition to his books and scientific articles Darwin wrote an autobiography giving his own opinions on the factors that had influenced his thinking. This was first published (with some sensitive passages on religious questions omitted) in the *Life and Letters* collected shortly after his death by his son Francis.[1] The complete text was published in 1958, edited by Nora Barlow.[2] The three volumes of the *Life and Letters* – the typical memorial to any great man of the Victorian era – was supplemented by a two-volumed *More Letters* in 1903. Several more volumes of letters were published before the modern project to publish the whole Darwin correspondence was begun. We now have available a *Calendar* listing 13,889 letters to and from Darwin, with a brief description of their contents.[3] Three volumes of the complete letters were available at the time of writing, covering the years to 1846; a fourth and fifth have since been published. Darwin's notebooks from the crucial years of his discovery in the late 1830s were originally published by Sir Gavin De Beer in the 1960s, but a splendid new edition is now available.[4]

These private papers allow historians to see behind the facade that Darwin and his family presented to the public. Even the autobiography – and Darwin's is honestly written by the standards of the genre – must to some extent be seen as an exercise in self-justification. We cannot altogether rely on Darwin's recollection of the steps by which he was led to his theory. After fifty years, no one could avoid an element of reconstruction in their memory of crucial events that shaped the course of their life. Equally significant is the distortion introduced by Darwin's family in their selection of the private material to be published in the posthumous *Life and Letters*. We have already noted that the autobiography itself was at first edited to remove some of Darwin's thoughts on the question of religion. The Darwin scholar Jon Hodge has recently written of what he calls the 'Franciscan view' of the Darwinian landscape – a view of Darwin's life and thought constructed by his son Francis through the choice of material to be published.[5] This tends to play down those aspects of Darwin's thinking that had not proved very fruitful, especially his lifelong interest in theories of 'generation' or reproduction where he favoured a model quite at variance with modern views on heredity. At the same time, the Franciscan view highlights those areas of his work that later biologists have accepted as important contributions, especially his studies of biogeography and the adaptation of organisms to their environment.

The notebooks and letters now available allow historians to go directly to the original sources instead of being forced to view their material through a filter imposed by an earlier selection process. What follows is an account of Darwin's career that tries to mediate between the need to reveal the scientific foundations of his theory and the necessity of including the insights generated by the latest scholarship. Hindsight is necessary to some extent if we are to identify those aspects of Darwin's work that turned out to have permanent value for biology. But hindsight is never enough to provide a sympathetic picture of a creative thinker at work in his cultural environment. Modern interpretations reveal the extent to which Darwin was absorbed in the scientific thinking of his time, as well as the extent to which he transcended that thinking to produce a radical new approach to the question of the origin of species. Equally significant is the emphasis placed by recent Darwin scholars on the interrelatedness of his various interests. The theory of evolution by natural selection did not emerge from a simple accumulation of facts or insights: it reflects Darwin's efforts to forge a meaningful relationship between a number of different interests and attitudes, all of which played a role in shaping the direction of his thoughts. Scientific and non-scientific factors both contributed to the creative synthesis which produced his theory.

The Darwin Family

Charles Robert Darwin was born on the 9th of February 1809 in Shrewsbury. His father, Robert Waring Darwin, was a successful and wealthy physician. Charles was the second son and the fifth of six children. His grandfather was Erasmus Darwin, also a physician, but internationally known for his poetic depictions of the natural world and his speculations about the nature and origin of life. Although Erasmus died several years before Charles was born, the younger Darwin certainly knew from an early age about the evolutionary ideas expressed in his grandfather's *Zoonomia*. Darwin's mother was Susannah Wedgwood, daughter of Josiah Wedgwood, whose pottery business was one of the early success stories of the industrial revolution. She died when he was only eight and he was raised instead by his elder sisters.

The Darwins were successful members of the bourgeoisie, and the family provided an intellectual and cultural environment that shaped Charles' attitudes throughout his life. In some respects the Darwins belonged to the more radical element of the liberal or Whig tradition. The men were free thinkers on matters of religion, having no interest in the platitudes of those who used Christianity to bolster the status quo. The women, however, were frequently more pious and more inclined to take the Bible seriously. Having gained their wealth through their own efforts, the Darwins had no patience with aristocratic privilege. They supported the free-enterprise system and expected it to yield unlimited progress as industrialization was extended through the economy. Having gained wealth, however, they expected to keep it and had no time for the kind of radicalism that we nowadays call 'socialism'. Charles was among the first generation of Darwins to draw upon the family fortune to avoid the necessity of earning a living. He was acutely conscious of the social position conferred by wealth and anxious not to do anything that would threaten the family's status.

Many of these characteristic family attitudes can be seen in the writings and activities of Erasmus Darwin.[6] He was a member of the Lunar Society which flourished in Birmingham from 1766 onwards. Here many of the founders of the industrial revolution, including James Watt, Matthew Boulton, Joseph Priestley and Josiah Wedgwood, met to discuss the role of science and invention in transforming society. Erasmus Darwin himself made a number of mechanical inventions. The Lunar Society actively encouraged social reform. They opposed slavery in the colonies and called for freedom of enterprise and freedom of expression at home. Like many radicals, they at first welcomed the French revolution and were vilified by George Canning and the reactionaries who attempted to stifle dissent in the face of the threat from across the English channel. Priestley's house was sacked by a mob almost certainly encouraged by government supporters, and Erasmus Darwin's poetry was lampooned in Canning's *Anti-Jacobin*.

Darwin's view of society was reflected in his view of nature. He was fascinated by the complex physical structures needed to maintain and perpetuate life. Charles Darwin would inherit this vision of nature as a complex mechanical contrivance – what one might almost call an engineer's view of living structures. But both Erasmus and Charles saw nature's powers as active and developmental: species were ever-changing entities within a complex network of ecological relationships, not the static products of a wise and benevolent God. They revered nature itself rather than its Creator, and although Charles would for a time become attracted to the more static view of design and adaptation expressed in Paley's *Natural Theology*, he would eventually return to Erasmus's more dynamic vision. To look for detailed anticipations of evolution by natural selection in Erasmus Darwin's poetry and in the *Zoonomia* is to miss the real point of the link between the two men. No doubt Erasmus did anticipate aspects of sexual selection and the struggle for existence, but the real influence he exerted lies not in a few isolated passages seeming to anticipate modern ideas but in a generally more dynamic view of natural relationships.

Erasmus Darwin saw sexual reproduction as the key to nature's creative activity. It was the generation of new individuals which gave life the power to respond to an ever-changing environment – a truly creative renewal allowing the species to gain from the purposeful activities of every generation of its members. Charles Darwin too was to share this fascination with reproduction. In the words of Jon Hodge, Charles Darwin was a 'lifelong generation theorist' who saw his theory of evolution as the expression of a mediation between individual repro-duction and the environment.[7] This is one of the less modern aspects of Charles Darwin's thought and it has its roots in Erasmus's views on the significance of sex as nature's great renovating power.

One consequence of Erasmus Darwin's beliefs was that he saw knowledge as something to be gained by an active interaction between the observer and the real world. True knowledge was not to be found lying ready-made in the Bible or the writings of classical antiquity. His eldest son died before the age of twenty from an infection contracted as a medical student at Edinburgh where he had been sent as an alternative to the classical education dispensed by the English univer-sities. Robert Waring Darwin, Charles' father, gained his medical training at Leiden. Thus when Charles was sent to Edinburgh he was following the family tradition in matters educational. His subsequent transferral to Cambridge introduced him to a more orthodox back-ground which at first attracted him and then made him extremely cautious when he began to realize that his own search for independent knowledge would lead him into conflict with received ideas.

Charles had a deep respect for his father, who survived until 1848. Robert Darwin was an immensely successful physician, thanks in part to his sympathetic attitude. He was also a good judge of character and a good businessman, amassing a fortune that allowed him to provide his children with an independent income when they grew up. Perhaps as a consequence of this success, Charles was later to recall that his father became 'very corpulent, so that he was the largest man whom I ever saw'.[8] Charles could hardly remember his mother, who died in July 1817. His brother, Erasmus, was interested in science but suffered from poor health throughout his life. Their sisters, Caroline, Susan and Emily, played an important role in the family, especially after the death of their mother. The sisters knew and loved the Bible and would thus have been a force inclining Darwin to a more conventional way of thought.[9] Charles' childhood cannot have been without trauma, yet on the whole he grew up in the bosom of a comfortable and well-integrated family. His desire to retain this kind of comfort and personal stability was to play a major role in determining his later choice of living conditions.

After a short spell at a local day school Darwin was sent as a boarder to Shrewsbury School, where the headmaster was Dr Samuel Butler (grandfather of the novelist Samuel Butler, with whom Charles was to come into conflict in later life). As a boarder he would have learned a certain amount of independence, yet he could visit his family regularly and would not have suffered the isolation typical of so many middle and upper class boys sent away to public school. Unfortunately, Dr Butler's style of education was severely classical, and Charles was later to confess that 'The school as a means of education to me was simply a blank'.[10] He had already developed a passion for shooting that was to survive into his university days, to be repudiated eventually as useless slaughter.

More seriously, he was also becoming interested in collecting minerals and in bird watching. Along with his brother he developed a passion for chemistry, and some of the earliest surviving Darwin letters discuss the purchase and use of chemical equipment. He was even rebuked by his headmaster for taking an interest in so unfashionable a subject. A recent study by Sylvan Schweber suggests that Darwin's chemistry may have helped to shape his view of the nature of living things.[11] Schweber points out that many of the texts that Darwin used encouraged the belief that nature is a system designed by God to work according to His laws and that living things are structures governed by these physical laws. To some extent, Darwin's later theorizing can be seen as an attempt to understand how these living structures interact with their environment in a law-bound manner over long periods of time.

University Life

Because Shrewsbury School was not a success, Darwin was sent to Edinburgh as a medical student at the early age of 16. The Edinburgh interlude is often passed over quickly by biographers, but historians are now beginning to suspect that he gained more from his time here than one might suppose. Edinburgh was then a centre for marine biology and Darwin was soon collecting sea creatures and subjecting them to dissection and microscopic observation. He became active in the Plinian Natural History Society. He also began to interact with Robert Grant, then regarded as one of the country's most promising young naturalists. We saw in chapter 2 that Darwin later professed to have been amazed to hear Grant speak out in support of Lamarck, but he can hardly have failed to appreciate that opinions considered highly unorthodox south of the border were circulating freely in Edinburgh.

In a detailed study of the Edinburgh period, Phillip Sloan has suggested that Darwin was already becoming an active naturalist with theoretical opinions and that interests aroused at Edinburgh were to remain important through into later episodes of his career.[12] Indeed, Sloan and other historians argue that we should abandon the old image of Darwin's thoughts undergoing a sequential development towards his mature views. Instead, they argue that certain themes and interests continued throughout his career, sometimes falling into the background only to be revived again later on. Inspired by Grant, he acquired an interest in colonial marine invertebrates such as the Hydrozoa and corals, then collectively known as 'zoophytes' (from the Greek for 'animal' and 'plant', indicating an animal that seems to grow like a plant). Grant himself believed that these creatures served as a bridge between the plant and animal kingdoms and that the reproduction of the zoophytes would throw light on the structure and functioning of both plant and animal bodies. Darwin at first opposed Grant's views and we know that he was less than enthusiastic about the transformist theory that Grant was building up in his researches. But the programme of research begun in Edinburgh would be revived on the *Beagle* voyage, at which point Darwin became more sympathetic to Grant's interpretation of the zoophytes' pivotal position between the two kingdoms.

Sloan argues that this later development was an important prelude to his adoption of transformist views, and points out that he was to retain a lifelong interest in the question of reproduction sparked off by this early research.

Darwin's official studies at Edinburgh were less promising. Apart from Dr Hope's chemistry lectures, he hated the courses he was required to take, although he was interested in the hospital rounds. In particular, he was bored by the lectures on geology offered by Robert Jameson, a prominent exponent of the Neptunist theory in which the whole earth was once supposed to have been covered by a vast ocean. The Neptunist approach emphasized mineralogy rather than the study of fossils. Jameson's views were already becoming outdated, but it seems to have been his boring style of lecturing that put Darwin off. He later recorded that the lectures were 'incredibly dull' and that 'The sole effect they produced on me was the determination never as long as I lived to read a book on Geology, or in any way to study the science'.[13] Jameson must have been a poor lecturer indeed to render Darwin so disgusted with a subject that he was later to take up with considerable enthusiasm. A more active disgust was engendered by some unpleasant experiences in the operating theatre, and before long Darwin was becoming convinced that he was not cut out for a medical career. Awareness of the fact that he need not work for his living must have encouraged him to look for alternative forms of education.

It was Darwin's father who suggested that he consider taking holy orders in the Church of England. At this time the clergy offered one of the few openings for those seeking a 'respectable' profession, apart from medicine and the law, and many clergymen were amateur naturalists. Darwin himself had some reservations about the 39 Articles of the Church and asked for time to consider, but he soon came round to the idea. As he later wrote: 'as I did not then in the least doubt the strict and literal truth of every word in the Bible, I soon persuaded myself that our Creed must be fully accepted'.[14] The use of the phrase 'persuaded myself' perhaps suggests some need to suppress doubt, and certainly one of Darwin's neighbours from Shrewsbury wrote that she was 'very much surprised' to hear that he had decided to become a DD rather than an MD.[15] To enter the Church Darwin had to take a degree from one of the English universities and he opted for Cambridge. Since he had now forgotten most of his classics he had to go to a private tutor for a few months before going up to Christ's College at the end of 1827.

At Cambridge Darwin encountered a very different atmosphere from that prevailing in Edinburgh, an atmosphere designed to reinforce, if only temporarily, the more conservative side of his character. Many of his fellow undergraduates intended to enter the Anglican church. He was to study classics, divinity and mathematics, but the natural sciences did not form a part of his curriculum. The Professor of Botany, John Henslow, gave public lectures which Darwin attended with enthusiasm, although he did not attend the equivalent lectures given by the Professor of Geology, Adam Sedgwick, and only began to take up an interest in geology towards the end of his time at the university. Henslow and Sedgwick, of course, were both ordained Anglicans who saw no conflict between their science and their religion because they interpreted nature as God's creation. In many respects, however, Cambridge was much less a bastion of High Church Toryism than Oxford. Indeed both Henslow and Sedgwick were Whigs who supported reform in the debates leading up to the passing of the Reform Bill in 1832. Darwin records that Henslow was the right-hand man to Lord Palmerston, who was to lose his seat as MP for the University because of his support for reform.[16] Sedgwick too backed Palmerston. The environment within which Darwin now functioned was thus theologically conservative but politically rather more in line with his family's liberal traditions.

The actual course work at Cambridge was of only marginal interest to Darwin and he was poor at mathematics. By working hard at his classics and divinity he managed eventually to gain a respectable degree; among the 'men who did not go in for honours' he was tenth in the list of January 1831. His autobiography records that the one topic that did interest him was the study of Paley's *Evidences of Christianity* and *Moral Philosophy*. He also studied Paley's *Natural Theology*, taking its basic assumption on trust and being charmed by its long list of adaptations showing how each species was designed by its Creator.[17] Paley reinforced his early interest in the adaptation of the living body to its environment, but provided a far more conventional explanation of the phenomenon. Eventually, of course, Darwin would be led back to considering the problem of the interaction between the organism and its surroundings in a less static context.

Darwin later claimed to have wasted much of his time at Cambridge on shooting and on having a good time with sporting companions. He also recollected with some puzzlement that he took an interest in music despite having no sense of tone or rhythm. His friends teased him by asking him to identify well-known tunes played at the wrong speed, under which circumstances even 'God Save the King' was unrecognizable to him. In fact, Darwin was far from idle in the field of natural history. Inspired by his second cousin W. Darwin Fox, he became an avid collector of insects (the *Correspondence* includes a long series of letters to Fox on this topic). It was Fox who introduced Darwin to Henslow, thus beginning a friendship which he was later to regard as one of the most influential of his whole life.

In effect it was Henslow who encouraged Darwin to become a full-time naturalist (one cannot say 'professional' naturalist since he did not earn his living from science, and the concept of a professional scientist was only just beginning to emerge at this time). Darwin attended Henslow's lectures and field trips. Soon he was going to the professor's weekly 'open house' and eventually he was invited to dine with the family on a regular basis. Henslow clearly saw that Darwin was a young man with a powerful interest in nature and did his best to encourage his scientific education. We must get away from the image of Darwin as the relatively untrained novice who set out on his voyage around the world with little or no understanding of the theoretical issues underlying natural history. Thanks to his Edinburgh experiences and his training under Henslow, Darwin was becoming as experienced as anyone of his age could have been expected to be. Darwin's interest in *scientific* natural history is clear from the enthusiasm with which he read Sir J. F. W. Herschel's *Preliminary Discourse on the Study of Natural Philosophy* of 1831, an influential statement on the problem of defining the scientific method.

One consequence of Darwin's growing enthusiasm was his ambition to study nature at its richest, in a tropical environment. He read and copied out long passages from the *Personal Narrative* in which Alexander von Humboldt described his voyages to South America around the turn of the century. Humboldt exerted a great influence on early

PLATE 2 Darwin in 1840. Portrait by George Richmond at Down House. (By permission of the Darwin Museum, Down House, courtesy of The Royal College of Surgeons of England)

nineteenth-century science through his assumption that the gathering and collating of information from all around the world would yield a new image of nature as a complex but integrated system. But it was Humboldt's own experiences in the tropics that fascinated Darwin. During his last year at Cambridge he conceived a plan to travel to the Canary Islands, possibly along with Henslow. This came to nothing, but Darwin's anxiety to travel in the cause of natural history ensured that he would jump at the chance of a voyage around the world.

After taking his examinations Darwin had to spend two further terms in residence at Cambridge and it was at this point in his career that he extended his interests to include geology. After his experience with Jameson at Edinburgh, he had avoided Adam Sedgwick's popular lectures and field trips at Cambridge. Now at last he began to see that geology had its own fascination. Sedgwick was building a reputation as one of the country's leading geologists and was just beginning the long series of investigations in North Wales that would lead to the establishment of the Cambrian system, then assumed to be the oldest system of fossil-bearing rocks.[18] He was a supporter of the catastrophist theory which assumed that the earth's crust had been shaped by periodic convulsions. Indeed, Sedgwick had only recently given up the belief that the last of these great catastrophes was a universal flood that could be identified with the biblical deluge. He would remain committed to the belief that God's hand could be seen at work in the history of the earth and its inhabitants, ultimately becoming a bitter opponent of Darwin's evolutionary theory.

At this stage, however, Sedgwick appealed to Darwin because his active approach to fieldwork demonstrated how to uncover the historical sequence in which the various formations of the earth's crust had been laid down. Darwin never forgot an episode in which Sedgwick dismissed an ancient fossil found in a gravel pit as an intrusion, pointing out that if it were genuine the find would upset the whole technique established for dating geological formations.[19] In the summer of 1831 Darwin accompanied Sedgwick on a trip to Wales and there can be no doubt that Sedgwick was by now taking his young companion seriously. A letter written to Darwin after the latter had returned home gives ample evidence of Sedgwick's confidence in his ability to follow the complexities of stratigraphy.[20] An earlier letter written by Darwin to Henslow suggests that to begin with he had absorbed his teacher's catastrophist opinions: 'As yet I have only indulged in hypotheses; but they are such powerful ones, that I suppose, if they were put in action but for one day, the whole world would come to an end'.[21] On the *Beagle* voyage this taste for discontinuity in nature's actions would soon be swept away under the influence of Charles Lyell's uniformitarianism.

Darwin had certainly not given up his desire to travel more widely and in August 1831 an opportunity at last emerged. Henslow was informed that Captain Robert Fitzroy was looking for a naturalist-companion to sail with him aboard HMS *Beagle* on a voyage to chart the coast of South America and the South Sea Islands.[22] On most naval voyages of exploration it was customary for the ship's doctor to study the natural history of the regions visited, but Fitzroy wanted someone who would be free from the crushing discipline that would govern his relationships with the rest of the crew. As First Lieutenant he had brought the *Beagle* home from her previous voyage after the captain had suffered a mental breakdown almost certainly induced by the isolation of command. A gentleman-naturalist would be someone with whom the captain could communicate on normal terms, and Fitzroy thus made it known that he would welcome aboard a man who had the right social and scientific qualifications. The hydrographer of the Navy, Captain Francis Beaufort, consulted his friend George Peacock of Trinity College, Cambridge, and the position was first offered to the Reverend Leonard Jenyns. But Jenyns felt unable to leave his parishioners, and it was at this point that Peacock turned to Henslow for advice.

Henslow immediately suggested that Darwin offer himself for the position. Socially, his background in the upper middle class was acceptable, and there can be little doubt that Henslow now knew Darwin to be a thoroughly competent naturalist despite his lack of formal training. Darwin was excited at the prospect of such a voyage but first had to obtain his father's permission. This proved difficult and it was not until Josiah Wedgwood II interceded on Charles' behalf that his father gave up his objections. A letter from Wedgwood to Robert Darwin dated 31 August reveals the objections that had to be answered.[23] Robert Darwin was concerned that the voyage would reflect badly on his son's character, especially if he eventually attempted to take up a position in the Church. In effect, he was worried that Charles might waste his time and energy on a project that would only make it more difficult for him to settle down in later life. Wedgwood responded that natural history was a perfectly respectable occupation for a clergyman but hinted that Charles's commitment to his studies opened up an alternative lifestyle for which the *Beagle* voyage would be a perfect preparation: 'His present pursuit of knowledge is in the same track as he would have to follow in the expedition.' The fact that Robert Darwin gave up his objections suggests that he reconciled himself to the prospect that his son now wished to devote his life to natural history: eventually Charles might hope to speak on equal terms with the leading scientific figures of the land, and this might give him the status and stability that would compensate for the disappointed hope that he would become a clergyman–naturalist. One of Darwin's fellow students at Cambridge who did in fact become a clergyman expressed exactly these sentiments to him: the voyage will give him the opportunity to be 'ranked with Brongniart, de Candolle, Henslow, Linnaeus & Co. – Whilst I, luckless wretch, am rusticating in a country Parsonage & shewing people a road I dont know – to Heaven'.[24]

There were still obstacles to be overcome. Having met Darwin, Fitzroy decided that the shape of his nose did not indicate sufficient firmness of character (he was a follower of the 'science' of physiognomy, according to which personality was reflected in the facial features).[25] Darwin himself seems to have had no doubts about Fitzroy, although they were to quarrel on occasions during the voyage. Eventually they came to an arrangement and Darwin moved aboard ship at Plymouth on 24 October 1831, although bad weather was to delay the departure until 27 December. Darwin had to arrange for everything he might need on a voyage originally intended to last three years, including guns, hand lens and microscope, equipment for geology and chemical analysis, and of course books. He took a number of reference works on natural history and the first volume of Charles Lyell's *Principles of Geology* – a work that Henslow advised him to read but on no account to believe. All this had to be crammed into an extremely small space: the *Beagle* was a specially adapted ten-gun brig of only 242 tons, with extra stores for the long journey and an augmented crew. Somehow Darwin got himself established aboard and waited miserably for the weather to moderate.

4

The Voyage of the Beagle

THE VOYAGE OF the *Beagle* is often presented as a watershed in Darwin's career, an experience which effectively converted him to evolutionism and thus shaped all his later thinking. Darwin himself encouraged this view when he wrote in the *Autobiography* 'The voyage of the *Beagle* has been by far the most important event in my life, and has determined my whole career.'[1] The story of how his discovery of the different finches inhabiting the Galapagos islands convinced him of the reality of transmutation has been told over and over again. The Galapagos themselves have become a focus of worldwide interest among naturalists, while Darwin's voyage has become one of the legendary foundations upon which modern scientific biology is supposed to have been built. Darwin's *Journal of Researches* based upon the voyage is his next most frequently reprinted book after the *Origin of Species* (and far ahead of the *Descent of Man*). Coffee-table books and even television series have been based on the voyage, exploiting the visual attractions of exotic locations in conjunction with the legend of scientific discovery.[2]

The scholars of the 'Darwin industry' have been forced to battle against the mythological character acquired by the voyage in order to reconstruct the true story. Many aspects of the traditional legend have

PLATE 3 HMS *Beagle* at Sydney Harbour in 1841. From a Watercolour by
Owen Stanley. (National Maritime Museum, Greenwich)

not been substantiated by the most recent scholarship. Thanks to the
work of Frank Sulloway, we now know that Darwin did not recognize
the significance of the Galapagos finches until after the *Beagle* had
departed. He had to use other people's collections in order to investigate
the problem of speciation in this unique environment. More generally,
the Darwin scholars have shown that we need to reinterpret our whole
picture of what he was up to during the five years he was away. The

voyage is traditionally interpreted with the benefit of hindsight: we know what use Darwin eventually made of his discoveries and we allow this to influence our evaluation of what he actually did. Once again, Darwin himself contributed to the problem by rewriting the later and more popular edition of the *Journal of Researches* to incorporate the fruits of his own reflections on the voyage's significance. Historians now argue that we must force ourselves to accept that his conversion to evolutionism came *after* his return to England. If we want to understand what Darwin was actually doing while circumnavigating the globe, we must look to the notebooks and letters written at the time – which reflect a very different set of interests to those imposed by hindsight.[3]

In a letter to Henslow written from Rio de Janeiro in May 1832, Darwin insisted that 'Geology & the invertebrate animals will be my chief object of pursuit through the whole voyage'.[4] To an extent that may surprise those bemused by the stories of fossil bones and Galapagos finches, this turned out to be an accurate prediction. In 1835 he wrote 'Since leaving Valparaiso, during this cruize [sic], I have done little excepting in Geology'.[5] To a large extent, Darwin built his reputation on the geological discoveries of the voyage: his views on the elevation of the Andes and on the formation of coral reefs by the subsidence of the underlying land surface were major contributions to the science which guaranteed him a position among the elite once they were published. In zoology too his later interests initially played second fiddle to his enthusiasm for topics in which he was already deeply involved before departure. As Phillip Sloan notes, the investigation of zoophytes (corals etc.) begun in Edinburgh was to remain a major focus of attention throughout the voyage, leading Darwin eventually to a new position on the relationship between the plant and animal kingdoms.[6] Darwin's initial failure to appreciate the significance of the Galapagos finches can easily be understood in the light of his preoccupations at the time. He was only gradually becoming aware of the problems posed by the geographical distribution of species and was not yet prepared to see birds and other vertebrates as important clues to the origin of species.

A computerized study of Darwin's letters home by Sulloway has confirmed the significance of geology in his thinking.[7] Sulloway has also been able to chart Darwin's changing attitudes on a number of topics in the course of the voyage. At first he saw himself very much as a half-trained amateur sent out to collect specimens for the experts back home. He frequently alluded to his ignorance in many areas of natural history and expressed fears that his collections would not be enough to guarantee him a hearing when he returned. He was deeply concerned by the lack of any acknowledgement from Henslow of the material he had sent home. Once he received letters from Henslow in 1834 he became more confident, realizing that it was the naturalists back home who were now dependent on him for specimens of newly discovered species.[8] It was in geology, however, that Darwin's confidence in his own powers increased most dramatically. After 1834 he began to advance theories on geological dynamics and he knew that he would be returning with exciting evidence that would influence theoretical debates in England. Sulloway argues that the most important result of the voyage was not the evidence it supplied for transmutation but Darwin's increased confidence in his own abilities as a scientific thinker, which encouraged him to tackle the deeper problem of the origin of species on his return.

Darwin left England as a partly trained naturalist with strong interests in invertebrate zoology and geology but with major gaps in his knowledge elsewhere. He returned with his original interests deepened and with a growing awareness of the interesting problems that might emerge from a study of biogeography in conjunction with his new commitment to geological uniformitarianism. Darwin himself published the geological results of the voyage, while he edited the detailed descriptions of the zoological specimens by Richard Owen, John Gould and other more experienced workers. The first edition of his general survey, the *Journal of Researches*, appeared in 1839, followed by a revised and more popular edition in 1845. As noted above, the latter version is still in print and the chapter numbers given below refer to this edition.

South America

The *Beagle* finally left England on 27 December 1831, after having twice been driven back by gales. Darwin was racked by seasickness, and it must have been doubly disappointing for him when the *Beagle* was refused landing at Tenerife because of fears that she might be carrying cholera. On 16 January 1832 they reached St Jago, the chief island in the Cape Verde group, and Darwin was able to go ashore (*Journal of Researches*, ch. 1). Shortly before landing he had collected fine dust which appeared to have been blown out to sea, perhaps from Africa. In the *Journal* he comments on the implications of this for the dispersion of the spores of cryptogamic plants – evidently an insertion inspired by his subsequent interest in the problem of species dispersal. (In chapter 8 Darwin notes that a grasshopper landed on the *Beagle* when the closest land not against the prevailing trade wind was the coast of Africa, 370 miles away.) The landscape of St Jago was barren, but Darwin made some useful geological observations. The *Beagle* next called at St Paul's Rocks, 540 miles from the South American coast. Here the only inhabitants were sea birds and the parasitic insects they hosted, a fact which gave Darwin food for thought on the question of how newly formed islands became populated.

On 29 February 1832 the *Beagle* reached Bahia or San Salvador in Brazil, allowing Darwin his first glimpse of a tropical forest.

> The day has passed delightfully. Delight itself, however, is a weak term to express the feelings of a naturalist who, for the first time, has wandered by himself in a Brazilian forest. The elegance of the grasses, the novelty of the parasitical plants, the beauty of the flowers, the glossy green of the foliage, but above all the general luxuriance of the vegetation, filled me with admiration.[9]

Most of April, May and June were spent in the vicinity of Rio de Janeiro (ch. 2). Darwin made several trips inland and did some work on invertebrates. He also enjoyed the social life of the town:

> Our chief amusement was riding about & admiring the Spanish ladies. –
> After watching one of these angels gliding down the streets; involuntarily

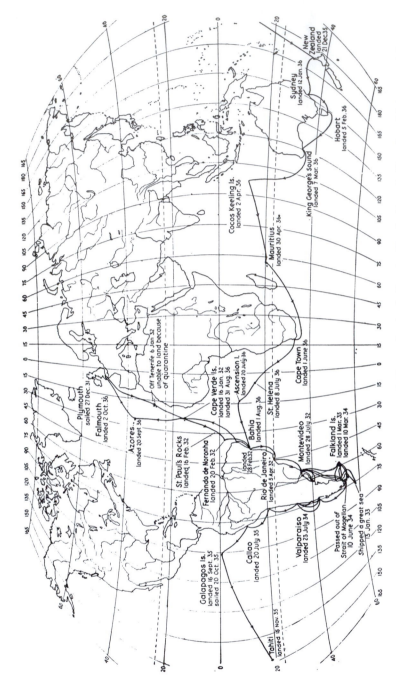

PLATE 4 Map of the *Beagle* voyage (1831–1836) From Gavin de Beer's *Charles Darwin* (London: Nelson, 1963)

we groaned out, 'how foolish English women are, they can neither walk
nor dress'. – And then how ugly Miss sounds after Signorita.[10]

More seriously, Darwin came face to face with a slave-owning society
and did not like what he saw. His family, of course, had been prominent
in the campaign to eliminate slavery from British possessions, and so
it is hardly surprising that Darwin hated the institution. The *Journal of
Researches* records several instances of the masters' cruelty to their slaves.
Darwin also notes a case in which an elderly female slave threw herself
from a cliff rather than allow herself to be recaptured: 'In a Roman
matron this would have been called a noble love of freedom; in a poor
negress it is mere brutal obstinacy.'[11] Slavery was here the subject of
Darwin's first quarrel with Fitzroy, who, as a staunch Tory, was inclined
to favour the institution. Darwin thought he might have to leave the
ship, but after letting off steam to his officers Fitzroy backed down.
The *Autobiography* also records another quarrel some years later at
Concepcion in Chile which ended in the same fashion.[12] Fortunately,
Darwin seems to have realized the strain under which Fitzroy was
working and was prepared to make allowances, except in the heat of
the moment. At least the two men did not quarrel over religion, since
Darwin at first retained his orthodoxy and only began to doubt the
moral authority of the Bible after his return to England.[13] Fitzroy would
later adopt a fundamentalist position on the question of the Genesis
story of creation, but at this point he was far more flexible in his beliefs.

In July 1832 the *Beagle* moved to Montevideo; she would spend much
of the next two years charting to the south while Darwin made frequent
trips inland. He travelled over the pampas or plains accompanied by
the gauchos, whose skills at riding and with the lasso and bolas he
records (chs 3 and 8). Sometimes they would stay with the local
landowners but often they slept out under the stars.

I am become quite a Gaucho, drink my Mattee & smoke my cigar, &
then lie down & sleep as comfortably with the Heavens for a Canopy
as in a feather bed. – It is such a fine healthy life, on horse back all day,
eating nothing but meat, & sleeping in a bracing air, one awakes as fresh
as a lark.[14]

Darwin obviously adapted well to the outdoor life and was quite prepared to adopt the living conditions of the countries he visited. In later years he would become a chronic invalid and something of a recluse, but as a young man he was certainly more adventurous.

In the course of his excursions Darwin made geological observations and collected both plant and animal specimens. He noted that the tucotuco, a rodent which had adapted to tunnelling like a mole, had become blind, a fact which he suspected would have delighted Lamarck (ch. 3). Later he found a viper which shook its tail like a rattlesnake although it had no rattles (ch. 5). The comments on this creature in the *Journal* make it clear that in the years immediately following his return to England Darwin saw this discovery as an indication of the intermediate stages by which unusual structures might have evolved.

The most important zoological discovery on the pampas, however, was a new species of rhea or South American ostrich (ch. 5). Darwin had soon become familiar with the common rhea but had been alerted by hearing the gauchos talk of a different form to the south. Even so, they had cooked and eaten a member of the new type before Darwin realized its significance. Fortunately, enough of the bird remained for him to send the specimen home, where the ornithologist John Gould named it *Rhea darwinii* in his honour. In later years the existence of the two species of rhea gave Darwin much food for thought. Their close relationship helped to establish the possibility that one might have been derived from the other by transmutation. But Darwin was also led to ponder the question of what determined the distribution of two such closely related forms. The newly discovered species had its main range to the south of the common rhea, but there was an intermediate territory occupied by both. This was one of the points which forced Darwin to rethink the conventional interpretation in which each species was supposed to be perfectly adapted to its home territory. Some difference in the conditions must define the two territories, but the two species are competing to occupy the intermediate zone and a slight change in the climate might allow one to expand its range at the expense of the other.

At several of the places he visited Darwin discovered the fossil bones of species that appeared to have become extinct in the fairly recent geological past. On an overland journey from Bahia Blanca to Buenos

Aires in September 1832 he found the remains of the giant ground sloth *Megatherium* along with fossil armadillos and a giant rodent, *Toxodon* (ch. 5). More specimens were found on a trip from Buenos Aires to Santa Fé a year later (ch. 7). At Port Desire in Patagonia he found a fossil which appeared to have affinities to the guanaco or llama, eventually to be named *Macrauncheria* (ch. 8). The comments on these discoveries in the *Journal of Researches* once again reveal how they had influenced Darwin's thinking in subsequent years. Noting that the rodent *Toxodon* also seemed to have affinities to the pachyderms, he observes: 'How wonderfully are the different Orders, at the present time so well separated, blended together in different points of the structure of the Toxodon!'[15] This was exactly the kind of relationship that convinced Darwin of the need to see evolution as a branching tree rather than a ladder of progress.

Macrauncheria turned out to be rather more complicated: it was named by the anatomist Richard Owen who originally saw it as an extinct relative of the llama and camel. Owen soon changed his mind, however, and accepted it as a pachyderm with only superficial camel-like features. The *Journal of Researches* suggests that Darwin had not entirely appreciated this point with its fatal implications for the assumption that *Macrauncheria* was a close evolutionary relative of the modern llama.[16] In fact it was at this point in the *Journal* that Darwin chose to discuss what became known as the 'law of succession of types', which highlighted the similarity of South American fossils to the continent's living species. The hint that such a relationship might imply transmutation can easily be read into Darwin's words by anyone familiar with his later theory: 'This wonderful relationship in the same continent between the dead and the living, will, I do not doubt, hereafter throw more light on the appearance of organic beings on our earth, and their disappearance from it, than any other class of facts.'[17] *Macrauncheria* may have been a disappointment to Darwin, but the fossil sloths and armadillos were obviously related to their modern equivalents. It would be wrong to suppose that Darwin saw the evolutionary implications of these fossils as soon as he discovered them, but clearly they were to influence his thinking enormously over the next few years.

On several of his journeys Darwin encountered the troops of General Rosas, who was engaged in a war of extermination against the native

Indians (chs 4 and 5). The gauchos and others of European extraction openly applauded Rosas' ruthless methods, leading Darwin to ask how this could happen in a supposedly Christian country. It is at least possible that this early experience of how one branch of the human race could exterminate another in order to take possession of its territory might have encouraged Darwin to visualize the relationships between species in competitive terms. Later on (ch. 7) Darwin became mixed up in one of the frequent revolutions that plagued the country and expressed the view that Rosas would soon become a dictator. He also allowed himself some caustic comments on the improvements that would have been made had the colonists been of British rather than Spanish origin.

In December 1832 the *Beagle* sailed south to the Straits of Magellan and the inhospitable island of Tierra del Fuego (ch. 10). On a previous visit to the island Fitzroy had brought away three natives, who had now been educated into the British way of life. In their native habitat the Fuegians seemed to epitomize the Europeans' image of the brutal and degraded savage – Darwin reports that they seemed willing to eat their old women rather than their dogs when food became scarce. Yet the three who had been brought away adapted well to European ways. Darwin was particularly attracted by one who had been named Jemmy Button, a lively character who was now most particular about the state of his dress. 'It seems yet wonderful to me, when I think over all his many good qualities, that he should have been of the same race, and doubtless partaken of the same character, with the miserable, degraded savages whom we first met here.' [18] All three were now to be returned to their homeland along with a missionary, Mr Matthews, in the hope that they would set up a civilized community. They were set ashore in Ponsonby Sound, where Jemmy Button found that he had forgotten his own language and seemed ashamed of his family. The sailors helped to build a wigwam and to dig a garden and the group were left with all the amenities of civilized life.

A few weeks later the *Beagle* called back at Ponsonby Sound, only to find Matthews already so alarmed by the behaviour of the other Fuegians that he asked to be taken away again. Many of the European goods had been pilfered and there had been threats of violence. Jemmy Button himself was not keen to stay on but was left behind when the

Beagle sailed back up the east coast of the continent. A year later she returned on her way through to Chile and the west coast. Fitzroy and Darwin were dismayed to find that Jemmy had now become a thin and haggard savage, just like his relatives. He came aboard but insisted that he wanted to stay with his people. As the *Beagle* stood out to sea, he lit a signal fire whose smoke seemed to bid her a 'last and long farewell'. The whole episode left Darwin with much food for thought concerning human nature in general and the relationship between the 'savage' and 'civilized' races in particular.

While the *Beagle* sailed the waters around the Falkland Islands and Tierra del Fuego, Darwin made numerous investigations of lower marine animals, especially the zoophytes or coral polyps. This work is acknowledged briefly in the *Journal of Researches* but is not described in full because Darwin felt the matter was of 'little general interest' (ch. 9). This is a good example of the extensive revision his notes underwent in order to produce a more popular text and to stress those aspects of his work that became important for his later theorizing. Nevertheless, he notes that his work threw much light on the reproduction of these 'plant-like animals' and helped to convince him that there was no sharp dividing line between the plant and animal kingdoms.

In the course of 1833 Darwin became increasingly anxious to hear from Henslow, to whom he had sent many of the specimens he had collected. In October he expressed his disappointment to his cousin, W. D. Fox:

> Excepting my own family I have very few correspondents; & hear little about my friends. – Henslow even has never written to me. I have sent several cargoes of Specimens & I know not whether one has arrived safely: it is indeed a mortification to me: if you should have happened to have heard whether any have arrived at Cambridge, do mention it to me. – It is disheartening work to labour with zeal & not even know whether I am going the right road.[19]

In fact Henslow had written in January and again in August, but it was the second letter which reached Darwin first while on the Falkland Islands in March 1834. The January letter, along with another written in December of the same year, did not reach Darwin until he was in Valparaiso, Chile, in July 1834.[20] Henslow was, in fact, extremely supportive: the *Megatherium* fossils had attracted wide attention, while the specimens of insects and plants were of great interest. Darwin's change of mood is obvious from his reply in July:

> Not having heard from you untill [sic] March of this year; I really began to think my collections were so poor, that you were puzzled what to say: the case is now quite on the opposite tack; for you are guilty of exciting all my vain feelings to a most comfortable pitch; if hard work will atone for these thoughts I vow it shall not be spared.[21]

From this point onwards he began to entertain serious thoughts of making a name for himself in natural history on his return. Henslow did, in fact, read some of Darwin's letters to the Cambridge Philosophical Society in November 1835 and arranged for them to be printed.

Early in 1834 the *Beagle* sailed through to the west coast of South America. Darwin explored the island of Chiloe and the coast around Valdivia (ch. 14). He was ashore near Valdivia on 20 February when he experienced a severe earthquake; the rolling motion of the ground was enough to make him feel giddy. On 4 March the *Beagle* sailed into the harbour of Concepcion to find that the town had been totally devastated by the earthquake. Darwin gives a graphic account of the human consequences of the destruction, but as a scientist his interest was aroused by the fact that the earthquake had produced a permanent elevation of the land surface. In Concepcion the elevation seemed to be about two or three feet, whilst thirty miles away beds of putrid mussels now stood ten feet above the new high tide mark. Exploring inland he found older beds of shells up to 1,000 feet high in the mountains; near Valparaiso they were found at an altitude of 1,300 feet. This linkage between the effects of a modern earthquake and the evidence of past geological elevation was to work a dramatic change in Darwin's views on geological dynamics.

During his exploration of the eastern coastal districts Darwin had already come across evidence suggesting that South America had been elevated from the sea. The *Journal of Researches* reports this as evidence of gradual elevation (chs 7 and 8) – yet we know from Darwin's papers that at the time he was still a supporter of Sedgwick's catastrophist theory. He had left England with the first volume of Charles Lyell's

Principles of Geology, which advocated the rival uniformitarian view in which all elevations and subsidences of the land were due to the accumulated effects of observable causes such as earthquakes and erosion. The second volume reached him in Montevideo in November 1832. Evidently Lyell's arguments took some time to convince Darwin, but the Concepcion earthquake played a decisive role. By July 1834 he was sure that the 1300-foot elevation shown near Valparaiso must be due to a succession of small earth movements, although he still supposed that the height of the Andes had been increased by only a small fraction during the Tertiary period.[22]

Darwin was now doing far more geology than zoology, thanks partly to expeditions into the great mountain chain of the Cordillera (ch. 15). By April of 1835 he was convinced that the whole range of the Andes had been elevated during the Tertiary period by the accumulated action of earthquakes.[23] Modern geological agents were indeed capable of producing major changes on the earth's surface. Darwin even speculated that the ruins of Indian buildings at heights where there was now no vegetation provided evidence that significant uplift had occurred during human history – although he conceded that destruction of water conduits might provide an alternative explanation of why the buildings had been abandoned (ch. 16). In August of that year he proclaimed himself 'a zealous disciple of Mr Lyell's views, as known in his admirable book'.[24] To a large extent, Darwin's later theorizing would be stimulated by his efforts to understand the implications of geological uniformitarianism for the geographical distribution of living species.

Across the Pacific

On 15 September 1835 the *Beagle* reached the Galapagos archipelago, a group of islands on the equator some hundreds of miles out into the Pacific Ocean from the South American mainland (ch. 17). The islands are of volcanic origin, much of the surface still being made up of lava covered only by stunted trees and brushwood. The *Beagle* called first at Chatham Island (Darwin uses the English names for the individual islands; the Galapagos now belong to Ecuador and have been given Spanish names). The air was often close and sultry with the sun beating down on the black rocks, leading Darwin to imagine that he might almost be in the infernal regions. There were craters and lava flows everywhere, giving clear evidence of recent volcanic activity. After a short stay the ship moved on to Charles Island where Darwin was able to explore at greater length. He climbed to the green uplands, the only part of the island receiving enough rain to support lush vegetation. The *Beagle* also called at Albermarle Island and James Island; on the latter Darwin was left for a few days while the ship went off to obtain water.

Darwin was particularly interested in the unusual animal life. He saw the giant tortoises which were often caught by sailors for food. One was so large that he could hardly lift it and he was told of others so heavy that even six men could not lift them off the ground. He also noted the marine iguanas, diving off the black rocks into the sea where they fed on seaweed. The birds were so unused to human visitors that they were almost tame. They could be killed with a stick or a hat, and Darwin records that he once pushed a hawk off the branch of a tree with the muzzle of his gun. He collected finches, mockingbirds, wrens and tyrant fly-catchers. It is, of course, the Galapagos finches that have come to symbolize the islands' influence on the development of Darwin's thought. According to the popular legend, he observed that there were a number of different species of finch, each adapted to a particular island, and realized almost immediately that they had been formed by divergent evolution. The true story of his discovery has turned out to be far more complicated.

The *Journal of Researches* makes little secret of the implications that Darwin was soon to read into the unusual fauna of the islands. They were geologically recent, yet had acquired a diverse and unusual population which nevertheless reflected a distinct American influence. The most surprising fact, Darwin notes, is that in some cases there is a recognizably distinct form on each of the islands. 'Hence, both in space and time, we seem to be brought somewhat nearer to that great fact – that mystery of mysteries – the first appearance of new beings on this earth.'[25] He goes on to explain that John Gould, the ornithologist to whom the finches were shown after he returned home, had recognized thirteen species divided into four groups. The beaks in particular showed significant differences, adapting the species to various modes of feeding. Darwin's comment is quite explicit: ' ... seeing this gradation and diversity of structure in one small, intimately related group of birds, one might really fancy that from an original paucity of birds in this archipelago, one species had been taken and modified for different ends'. Darwin had come to realize that sample populations derived from a South American ancestor had diversified on the isolated islands, each population adapting in its own way, until eventually a group of related species had been formed.

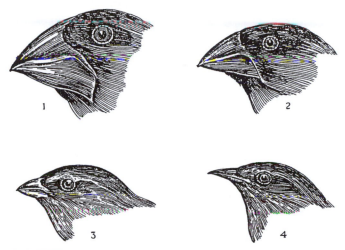

PLATE 5 Heads of four of the Galapagos species of ground finches illustrating the variation in beak structure. From Darwin, *Journal of Researches*, ch. 17

The analysis of Darwin's notes by Sulloway has demonstrated the extent to which the *Journal* account was a product of hindsight.[26] There was no 'eureka' experience on the Galapagos and Darwin's recognition of their significance developed only slowly over the next few years. In fact Darwin left the Galapagos without realizing their full significance: he did not collect any of the giant tortoises, while his collection of finches was incomplete and did not identify the islands from which the individual specimens had been taken. Darwin had to borrow specimens from other collections in order to test the conclusions he began to reach after his return to England. The mockingbirds may have offered him a more immediate clue than the finches, even though it is the finches which have come to be regarded as a paradigmatic example of island speciation. In fact the distribution of the finches is rather more complex than the popular legend allows since there has been a certain amount of migration from one island to another.

The *Journal* does admit that Darwin was slow to see the significance of the differences between the populations of the various islands.

> My attention was called to this fact by the Vice-Governor, Mr. Lawson, declaring that the tortoises differed from the different islands, and that he could with certainty tell from which island any one was brought. I did not for some time pay sufficient attention to this statement, and I had already partially mingled together the collections from two of the islands. I never dreamed that islands, about fifty or sixty miles apart, and most of them in sight of each other, formed of precisely the same rocks, placed under a quite similar climate, would have been differently tenanted.[27]

The *Beagle* sailed away before Darwin had time to exploit Lawson's information. It was on the long journey home, while he was writing up his ornithological notes, that he first began to suspect that the evidence from island groups might upset the accepted view on the stability of species. Only after showing his specimens to Gould on his return to England did he appreciate the full implications of the facts that he had so nearly missed seeing altogether.

The *Beagle* now sailed across the Pacific, calling briefly at Tahiti and New Zealand (ch. 18). Darwin did not like New Zealand: the natives lacked the charm of the Tahitians, while the white settlers were 'the refuse of society'. He was more impressed when they called at Sydney, Australia, a 'testimony to the power of the British nation' (ch. 19). Here Darwin was able to take a trip inland. He thought the Aborigines were less degraded than the Fuegians but commented on the 'mysterious agency' which seemed to afflict natives everywhere when their land was penetrated by Europeans, often leading to their near extinction. After a brief visit to Tasmania and to King George's Sound on the southwest corner of Australia, the *Beagle* set out across the Indian Ocean.

At Keeling Island in the Cocos group, Darwin observed coral reefs and the minute creatures responsible for them (ch. 20). We know that these studies helped to complete his long programme of investigation into the reproduction of these 'zoophytes', convincing him that there was no sharp distinction between the plant and animal kingdoms. But the *Journal of Researches* concentrates on his theory to explain the origin of coral reefs. Darwin had conceived this theory before he had even had the chance to examine a reef firsthand. Having recognized that South America was undergoing a gradual elevation, he realized that the land underlying the South Seas might be subsiding equally gradually. Since the coral animals can only live in shallow water, the subsidence would explain how the reefs were formed. As an island gradually sank into the ocean the corals would continually build up to the surface of the ocean, forming a fringe around the island and – when the original surface at last disappeared altogether – an atoll. This theory, published also in a separate volume on his return, was an important product of his conversion to Lyell's uniformitarianism and a major influence on Darwin's growing confidence in his powers as a scientific thinker.

The *Beagle* now made her way home via a circuitous route, calling at Mauritius, the Cape of Good Hope, St Helena and once again Bahia in Brazil (to complete her chronometric observations). From St Helena Darwin wrote to Henslow asking him to put his name forward for membership of the Geological Society of London.[28] Sedgwick had, in fact, read some of Darwin's letters to Henslow at the Geological Society where they had generated considerable excitement. Not surprisingly Lyell was anxious to make Darwin's acquaintance. A letter from Darwin to his sister Caroline makes it clear that in these last months of the voyage he had picked out for himself a career not as a country vicar but as an active member of the London scientific community.[29] He knew that he had made important discoveries in geology (more so, at this point, than in zoology) and looked forward with confidence to his participation in the scientific debates of the capital. He welcomed a last brief chance to see the tropical scenery of South America but thanked God he would never again have to visit a slave country (ch. 21). The *Beagle* made one last passage across the Atlantic, arriving in England on 2 October 1836. Darwin left the ship at Falmouth and set out at once to visit his family.

5

The Crucial Years:
London 1837 – 1842

L EAVING THE *Beagle* at Falmouth, Darwin travelled night and day to Shrewsbury to visit his family. In the days before the advent of the telegraph, one can imagine the joy with which he was greeted when he arrived unannounced just before breakfast on 5 October 1836.[1] He was relieved to find his family all well but was forced to depart all too soon for London to superintend the removal of his things from the *Beagle*. He then decided to spend a few months in Cambridge but soon saw that his future lay with the scientific elite of the capital. He moved to London in March 1837 and lived there until 1842, when he bought the house at Down in the Kent countryside where he spent the rest of his life. The move to Kent was to a large extent precipitated by the chronic illness which now threatened to flare up whenever he became too excited. In the meantime his personal circumstances had changed completely. He married his cousin, Emma Wedgwood, in January 1839 and their first child was born at the end of that year.

The London years were Darwin's one period of exposure to a really active scientific community. In a study of this period Martin Rudwick has argued that it can be no coincidence that it produced Darwin's great theoretical breakthrough in biology.[2] His public reception as a geologist who could talk on equal terms with all the great men in the science reinforced the growing confidence of the *Beagle* years. He could now turn – in private, of course – to the problem of the origin of new species which had begun to intrude on his thoughts as he reflected on the implications of his biogeographical observations. In July of 1837 he opened up his first notebook on the transmutation of species, beginning the process that would lead eventually to the theory of natural selection. The first reasonably comprehensive outline of that theory was written in 1842, just before the move to Down. Historians are still arguing over the thought processes that led to the discovery and over the factors, both scientific and non-scientific, which may have influenced his thinking.

Man about Town

Darwin had his *Beagle* specimens sent to Cambridge, where he had determined to spend a few months sorting them out. He stayed briefly with Henslow but then moved into lodgings. He found the social life of Cambridge very congenial and stayed longer than he had originally intended. Nevertheless, he knew that London must be his final goal if he was to participate in the important geological debates of the time. Darwin was not a natural city-dweller but he was prepared to make the sacrifice in order to establish his name as a scientist: 'I grieve to find how many things make me see the necessity of living for some time in this dirty odious London.'[3] He moved to London in March 1837, staying briefly with his brother Erasmus and then taking lodgings a few doors away at 36 Great Marlborough Street.

Darwin was concerned about the fate of his specimens and at first complained about the disinclination of other naturalists to take an interest in them. The exceptions were Owen, who was anxious to dissect the preserved animals, and Grant, who agreed to look at the corals.[4] The fossil bones eventually found a home with Owen at the Royal College of Surgeons. Publication of the zoology of the *Beagle* voyage soon began to move ahead much faster than Darwin had at first anticipated. Five parts were eventually issued between 1838 and 1842: 'Birds' by John Gould, 'Fish' by Leonard Jenyns, 'Fossil Mammalia' by Owen, 'Mammalia' by G. R. Waterhouse and 'Reptiles' by Thomas Bell. As the editor, Darwin was encouraged to petition the government for a grant to aid the cost of publication, necessary because of the expense of preparing illustrations. After some negotiation he obtained the sum of £1,000.[5] At first he had intended to include descriptions of the invertebrates in this series but the grant ran out before this could be done. Darwin himself described some of the invertebrate species but many were sent out to other naturalists who published descriptions in individual papers. This activity brought him into contact with many of the specialists working in the country, thus establishing the foundations of the communications network by which he would eventually gather information for his species research. At the same time, of course, Darwin was writing up the more general *Journal of Researches*, the first edition of which appeared in 1839. This was his attempt to follow in the footsteps of his early hero Humboldt, providing a 'personal narrative' of travel experiences along with scientific observations.

He was also deeply involved with the publication of his geological observations and theories. Three volumes, on coral reefs, South America, and volcanic islands, appeared between 1842 and 1846. The first and most important was *The Structure and Distribution of Coral Reefs* which took Darwin twenty months of hard labour but advanced his own highly original theory of reef formation. This book was very much a product of the London years and symbolized Darwin's emergence as a respected geological theorist. He was now deeply involved with the Geological Society of London which – as Rudwick points out – was then the most active scientific society in the country.[6] Although ostensibly committed to the collection of factual data, the society's papers often alluded to the confrontation over geological dynamics between the catastrophists (of whom Sedgwick was still a leading member) and Lyell's uniformitarians. Open debates took place at the meetings after the reading of formal papers. As a supporter of Lyell, Darwin was soon acknowledged as one of the elite group entitled to speak with authority on the most basic theoretical issues.

Volume 2 of the *Correspondence* includes a number of items showing that the Geological Society already operated the modern system of having papers refereed before publication. Darwin himself reviewed the papers of others, while his own submissions received the same treatment.[7] Perhaps his most important paper was an attempt to explain the famous 'parallel roads' that run along the sides of Glen Roy in Scotland. Darwin had seen similar formations in South America and assumed that they were shorelines formed when the land had been partially submerged beneath the sea.[8] He was subsequently forced to abandon this explanation in favour of an alternative suggested by Louis Agassiz, who invoked the concept of the Ice Age and argued that Glen Roy had become a lake when its mouth was blocked by a glacier. Although Darwin dismissed the paper as a 'great failure', it was in fact a useful exercise encouraging him to link diverse facts together.[9] When first published it was greeted as a serious contribution, helping to confirm Darwin's position as a leading member of the Society. Already by March 1837 he had turned down an offer to become one of the Society's Secretaries, a decision he later explained to Henslow by citing pressure of work and his aversion to abstracting papers for publication.[10] He did, in fact, accept this position in February 1838 and was elected Vice-President in 1843. He was elected a Fellow of the Royal Society in January 1839.

As an active member of the Geological Society, Darwin interacted regularly with some of the finest scientific minds in the country. Not surprisingly he was soon on good terms with Lyell, whom he praised for his clear-mindedness and his sympathy with the work of others.[11] In a letter to Lyell in 1841 discussing his Glen Roy theory, Darwin wrote: 'it is the *greatest* pleasure to me to write or talk Geolog. with you'.[12] He saw a great deal of Owen, who was describing the *Beagle* fossils. Darwin confessed that he was unable to understand Owen's personality and other friends warned him that Owen was a difficult man to get along with (in subsequent years he became a bitter opponent of the selection theory, although not of evolutionism in general). Sir J. F. W. Herschel and William Whewell were important contacts because they were recognized authorities on the scientific method, and Darwin

PLATE 6 Charles Lyell. From *The Life, Letters and Journals of Sir Charles Lyell* (London, 1881), vol. II, frontispiece

was anxious to ensure that his own theorizing – both public and private – met the accepted standards of the time.[13] He often talked with the great botanist Robert Brown but never on theoretical questions. These interactions took place as much at informal dinners as they did in the formal atmosphere of the scientific societies. Darwin also attended the famous evening parties given by the mathematician Charles Babbage. Although he occasionally met literary figures such as Macaulay and Carlyle, it was the scientific contacts that were the real heart of his London experience.

Enthusiastic as he was about his science, Darwin was becoming increasingly aware of the limitations to his personal life that would follow from a life devoted to nothing but work. A note pencilled on the back of a letter, probably in April 1838, discusses the options open to him and raises the possibility of marriage. Then in July of the same year he drew up a table of the advantages and disadvantages of marriage.[14] The disadvantages included the loss of freedom and of time to pursue scientific work but the advantages were now becoming all the more obvious to him.

> My God, it is intolerable to think of spending ones whole life, like a neuter bee, working, working, & nothing after all. – No, no won't do. – Imagine living all one's day solitarily in smoky dirty London House. – Only picture to yourself a nice soft wife on a sofa with good fire, & books & music perhaps – Compare this vision with the dingy reality of Grt. Marlboro' St.
> Marry – Mary[sic] – Marry Q.E.D.

His father seems to have advised him not to delay too long if he intended to have children, and in November of that year he proposed to and was accepted by his cousin, Emma Wedgwood. They were married on 29 January 1839 and set up house at 12 Upper Gower Street in London.

Darwin may have begun his search for a bride in a somewhat cold-blooded manner but there can be no doubt of his deep affection for Emma. She became the mainstay of his emotional life, even more so after the move to the relative isolation of Down. Their first child, William Erasmus, was born within a year, and altogether Emma bore him ten children. Darwin became very much a family man, with Emma superintending the running of the household in the typical Victorian manner. At the same time he was very much aware of their differences over matters of religion. Emma was a sincerely religious woman; she went to church regularly, had the children baptized and read the Bible with them. Darwin had been warned not to reveal his increasingly unorthodox views on religion to his wife but it is clear that he felt obliged to open his heart to her. He must have done so at a fairly early stage in the relationship because already in 1839 Emma wrote him a letter outlining her misgivings about the direction of his thoughts. She obviously respected his decision to be honest with her but was disturbed that his scientific theories were leading him to neglect the need for faith. At the bottom of her letter Darwin himself added these words: 'When I am dead, know that many times, I have kissed & cryed over this.'[15] He remained acutely aware of the dismay his ideas evoked in the woman he loved, and the fact that their marriage remained such a comfort to him suggests how strong their mutual attraction must have been to overcome this intellectual barrier. At a more practical level, Emma provided Darwin with an ever-present reminder of the difficulties he would experience in trying to present his views to his more orthodox colleagues and to the public at large.

Even before his marriage Darwin had begun to experience periods of illness. He had palpitations of the heart, headaches and an upset stomach, the attacks often being brought on by spells of hard intellectual work or any kind of excitement. This was in fact one of the reasons he originally gave for turning down the position of Secretary to the Geological Society.[16] In the end he was forced to leave London for the isolation of Down in order to escape the pressures that tended to precipitate his attacks. In later years he frequently took the 'water cure' at various hydropathic establishments, seldom with any long-lasting effects.

Much ink has been spilled over the question of Darwin's illness and a variety of retrospective diagnoses have been offered.[17] Many writers feel that the symptoms were the product of a psychological disturbance. At one time it was fashionable for analytical psychologists to postulate a neurotic origin for the symptoms, perhaps arising from an unconscious hatred of his overbearing father. Another popular suggestion was that

PLATE 7 Emma Darwin in 1840. From *Emma Darwin: A Centenary of Family Letters* (London, 1915), vol. I, frontispiece. (Reproduced by kind permission of John Murray Publishers and the Bodleian Library, Oxford)

he had contracted the parasitic infection known as Chagas' disease after having been bitten by the 'great black bug of the Pampas' while on the voyage of the *Beagle*. One problem with this interpretation is that Darwin complained of heart palpitations even before setting out on the voyage. He may well have had a congenital weakness of the stomach and nervous system, and current thinking seems to favour the view that psychological stress played a major role in precipitating his attacks. These attacks could well have become more intense after he began to realize that his views on the species question would bring him into conflict with public opinion and, more directly, with his wife's religious feelings. The move to Down in 1842 would thus have been a direct result of the tensions created by the development of the theory of natural selection in the late 1830s.

The Origins of the Selection Theory

Although primarily working in geology, Darwin became increasingly interested in the problem of reconciling Lyell's uniformitarianism with the geographical distribution of species. His 'Red Notebook', begun at the end of the *Beagle* voyage and finished in June 1837, reveals his first speculations on transmutation. He then opened up a series of notebooks known as the A, B and C Notebooks; A is still concerned mainly with geology, but B and C are addressed directly to the problem of transmutation and its implications. After July 1838 the notes are divided into two categories: the D and E Notebooks deal with biological evolutionism, while the M and N Notebooks concentrate on the potential implications that such a theory must have for mankind. Only in the E Notebook did the final version of the selection theory appear. The whole series thus provides a detailed record of Darwin's thoughts as he struggled to create a theory that would explain the facts at his disposal. The notebooks were originally made public through the work of Sir Gavin De Beer and Sydney Smith. A fine modern edition has now made this unique record of scientific creativity available to scholars everywhere (references to the *Notebooks* below are to the original pagination, clearly indicated in this edition).

The availability of Darwin's papers has allowed a veritable 'Darwin industry' to emerge within the history of science. Over the past few decades the origins of his theory have been examined in minute detail and from a variety of perspectives. In 1984 an article which merely attempted to survey the available literature on the topic ran to nearly fifty pages.[18] Since then a number of important studies have been published, including the authoritative series of papers edited by David Kohn in *The Darwinian Heritage*. A general survey of Darwin's life cannot afford to become bogged down in the technical details that form the basis of the Darwin industry's debates, but we must give at least an outline of the consensus that is slowly emerging on the interpretation of this most important step in his thinking. It is now clear that there was no 'eureka' experience in which Darwin was converted to evolutionism or in which he suddenly conceived the mechanism of natural selection. His ideas underwent a process of continuous development passing through phases in which he tried out various ideas that had to be modified or even abandoned as he broadened the range of his thinking. It has also become clear that both scientific and non-scientific factors played a role in shaping his thoughts. Natural selection was not a simple induction from observed facts, but nor was it a mere reflection of the competitive ethos of Victorian capitalism. Darwin drew upon a whole range of influences and synthesized them to give a unique explanatory model for the origin of species.

The speculations recorded in the species notebooks were not, of course, made public at the time, since Darwin knew that he was now investigating a topic with radical implications. Nevertheless, some historians have stressed that his species work needs to be interpreted in the light of his public activities as a geologist and a naturalist.[19] Lyell's uniformitarianism was both a key influence on his thoughts about species and the basis for his published ideas on coral reefs and the 'parallel roads' of Glen Roy. Darwin's coordination of the process by which the *Beagle* specimens were described and classified was also linked directly to his adoption of the transmutationist hypothesis. It was the ornithologist John Gould who made it clear to him that the Galapagos finches were distinct but closely related species and thus precipitated his doubts about creationism. His interaction with Owen on the fossils also bore directly upon his thinking about the relationships between species in time. Nor did Darwin conceal the fact that he was becoming interested in the question of evolutionism. In November 1839 he wrote to Henslow: 'I keep on steadily collecting every sort of fact, which may throw light on the origin & variation of species.'[20] Lyell and several other naturalists were also aware of his activities. By 1843 he was corresponding with G. R. Waterhouse, the zoologist who had described the mammals discovered on the *Beagle* voyage, in order to test his ideas on the implications of evolutionism for classification.[21]

These professional contacts were supplemented by an extensive reading programme and by direct appeals to animal breeders. Darwin read systematically in the areas of biology that he thought might bear on his research and in works on psychology, political economy and other topics that would throw light on the origins of human characteristics (see the list of books at the end of the B Notebook). In 1839 he issued a printed list of questions to be answered by breeders.[22] Nevertheless, it is clear that despite his wide search for information he was gradually moving towards a position that was uniquely his own.

The question of Darwin's originality – or lack of it – has long exercised the minds of historians. Some have supposed that he merely connected together a fairly obvious set of observed facts in order to synthesize the idea of natural selection. Others have treated the mechanism as little more than a projection onto nature of the competitive individualism of his social background. A comparison of his ideas with those suggested by other naturalists at the time soon reveals, however, that he used his scientific and cultural resources in a highly creative way to generate a theory that went far beyond anything that his contemporaries could envisage.[23] There was certainly a growing willingness to admit that nature was a scene of constant struggle and a few naturalists were able to see that unfit individuals might be eliminated. But it is one thing to imagine a species being 'purified' through the wiping out of deviant individuals, quite another to see that the selection of random variations might actually change the character of a species. There were some writers who saw that struggle might be the basis of a creative process but did not formulate the theory of natural selection. The philosopher Herbert Spencer is a good example: he saw the struggle for existence as a stimulus to a Lamarckian evolutionary process. To say that Darwinism was somehow 'in the air' obscures the subtle interaction of social and scientific factors that inspired Darwin's efforts and the highly original nature of his solution to the problem.

Darwin became determined to extend Lyell's methodology of

explaining past changes in terms of observable causes into the one area from which Lyell himself had excluded it.[24] Where Lyell rejected Lamarck's naturalistic theory of transmutation, Darwin accepted the basic idea of evolution and began to search for an alternative mechanism. The interpretation of his *Beagle* discoveries, especially from the Galapagos, was crucial to this decision. His first mention of transmutation (Red Notebook, pp. 127–33) coincided with John Gould's announcement that the Galapagos mockingbirds were distinct species. Yet, as Frank Sulloway points out, no one else regarded the Galapagos species as proof of evolution, not even Gould himself.[25] It was Darwin alone who decided that the resemblances between the species he had discovered in the Galapagos and in South America were so close that they must imply community of descent. From the start his model of evolution was based on the assumption that many different families evolved independently from one another – there could be no simple ladder of progress that all must ascend. The link in space between the common rhea and his own newly discovered species was identical with the link in time between the extinct *Macraucheria* and the modern llama. At this point he seems to have thought that the generation of one species from another must take place suddenly by a leap or saltation. He also entertained the possibility that species, like individuals, are born with a fixed lifespan after which they must inevitably decline to extinction.

The B Notebook launches his main investigation into the problem of a mechanism for transmutation, defining a conceptual scheme that was to remain largely intact through into the E Notebook where the new idea of selection finally emerged. Darwin heads his notes 'Zoonomia' to make clear his intention of following Erasmus Darwin's quest for the laws governing the organic world. As Jon Hodge has emphasized, a leading feature of his search was the phenomenon of 'generation' or sexual reproduction.[26] His views in this area were distinctly pre-genetical in character and they remained an integral part of his thinking throughout the rest of his career. Darwin would eventually build his theory of natural selection on his conviction that evolution must be a process of mediation between the environment and the sexual reproduction of individuals that maintains the population. Indeed he held that the purpose of sex was to ensure the creation and preservation of the variations that allow the population to respond to a changing environment. 'Why is life short. Why such high object generation. – We *know* world subject to cycle of change, temperature & all circumstances which influence living beings' (B, pp. 2–3). The father's role in reproduction was to ensure that the child was not merely a 'bud' from the mother's tissues (a natural clone, in modern terminology). Variations occurred when the growing organism was exposed to new conditions which tended to upset or distort its normal path to maturity.

It is important to note that Darwin's early thinking – including his original formulations of the selection theory – was distinctly teleological in character. He still believed that God had instituted the laws governing reproduction to maintain species in a state of perfect adaptation to their environment. A changing environment actually stimulates the production of the variations necessary for evolution. To begin with, however, he seems to have thought that the variations would automatically be adaptive – a form of Lamarckism. Only after his full appreciation of the struggle for existence did he come to believe that a changed environment disturbs growth to produce random or undirected variation. Even then he continued to accept that sexual reproduction was the Creator's way of ensuring that the raw material for adaptive evolution is always available when required.

In other respects, though, Darwin was already making a radical break with contemporary thinking. He assumed that new species would only appear when samples drawn from the old population were geographically isolated under new conditions. With the Galapagos in mind, he immediately realized that this would entail a branching open-ended model of evolution. The B Notebook already contains diagrams depicting a tree or branching 'coral' to illustrate the relationship between species (B, p. 26 – the coral is actually the better analogy since only the ends of the branches are still alive). Such a model was a far cry from the linear scale of the Lamarckians or of Chambers' *Vestiges* and Darwin realized that one of its most powerful consequences was its ability to explain the grouping on which biological classification is based. Similar species share a common ancestor, but where there is no obvious connecting link alive today we can assume that the intermediate species have been eliminated by extinction.

One of the most important characteristics of Darwin's approach is

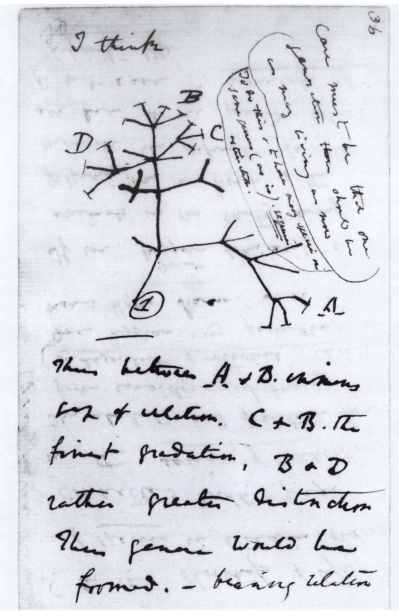

PLATE 8 Diagram to illustrate branching evolution; from Darwin's Notebook B, p. 36. (By permission of the Syndics of Cambridge University Library)

that it rejects the ambition to reconstruct a detailed history of the progress of life on earth. Knowledge of the process of adaptive evolution helps us to understand the relationships between living forms and throws some light on the fossil record. But the process is irregular and unpredictable because we can never hope to understand all the possibilities for migration that may open up for a species in the course of its history. The fossils themselves represent so small a proportion of all the species that have lived in the past that it is quite impossible to reconstruct a detailed evolutionary ancestry for every modern species. The unique nature of Darwin's approach was that it offered an understanding of historical processes in terms of observable causes *without* requiring him to attempt the impossible task of describing the whole history of life on earth. Evolution helps us to understand the nature of life on earth through its past – but only in the sense that it illuminates the general processes that have given rise to the particular living species we see today.

Although he accepted that some evolution must be progressive in order to explain the appearance of higher forms of life, Darwin had already realized that there could be no preordained pattern of ascent and that much adaptive evolution was not progressive in any meaningful sense of the term. Indeed the B Notebook contains one of his earliest protests against the simple assumption of a progressive scale that would continue to underlie much nineteenth-century evolutionism. 'It is absurd to talk of one animal being higher than another. – *We* consider those where the {cerebral structure/intellectual faculties} most developed, as highest. – A bee doubtless would when the instincts were' (B, p. 74). Here Darwin expresses the view that in a branching system comparison of one branch with another must inevitably be a subjective process. Progress must occur, but it is very difficult to define, and certainly not an inevitable consequence of all evolution. Thus Darwin rejected the developmental model that would be preferred by most of his later 'followers'.

Having decided to search for a natural mechanism of organic change, the C Notebook shows Darwin turning to the work of the animal breeders in the hope that it would throw light on the problem.[27] This was a unique move which played a vital role in determining the direction of his thoughts. No other naturalist saw the significance of artificial selection – not even Alfred Russel Wallace, the naturalist who independently discovered a form of natural selection two decades later. Of all the naturalists working in the 1830s and 1840s, it was Darwin alone who put together a theory of divergent evolution based on the natural selection of random variation.

In his autobiography Darwin implied that he was led to the idea of natural selection by observing the activity of animal breeders. The *Origin of Species* would later use a discussion of artificial selection as a means of introducing the reader to the theory of natural selection. But the notebooks reveal that artificial selection did not in fact provide the crucial insight. Darwin's memory of his discovery must have been distorted by his later decision to use artificial selection as a model. At first he studied the breeders' work for clues without accepting that there was any direct connection between the production of domesticated varieties and the evolution of species in nature. He was aware that the breeders worked by selection or 'picking' but saw no reason to use this as an analogy for natural change. Nor did he at first believe that the random or undirected variation from which breeders select was of any significance. Throughout the C and D Notebooks he remained committed to the view that adaptive variations must somehow be elicited automatically in a changed environment.

It was toward the end of the D Notebook in September 1838 that Darwin first begins to refer to Malthus and the principle of population. He had read T. R. Malthus' *Essay on the Principle of Population* not (as the autobiography implies) for amusement but as part of his systematic reading programme devoted to the human implications of evolutionism. There can be little doubt that in the end the concept of the 'struggle for existence' described by Malthus played a major role in switching his thoughts onto the path that led towards natural selection. At the beginning of the D Notebook Darwin later wrote, 'Towards close I first thought of selection owing to struggle'. Yet modern analysis of the Notebook suggests that Darwin's first use of the population principle was merely to modify his existing conceptual scheme. A passage in which he likens nature to a multitude of wedges being driven against one another (D, p. 135e) seems to refer only to competition between species, each trying to outbreed its rivals and take over their territory.

Only in the E Notebook does Darwin finally realize that the constant pressure created by the tendency for population expansion must entail a struggle between the individual members of the same species. Now at last the mechanism of natural selection begins to take shape. Darwin realized that the random or undirected variation from which the breeders select to create their varieties must also exist in nature. He was inclined to think that there would be less variation in nature because he saw changed conditions as the factor that would stimulate the production of new characters. Domesticated species were kept in unnatural conditions and thus exhibited considerable variability – but even in nature the gradual environmental changes produced by geological forces must occasionally generate some individual variations. Given the ever-present struggle for existence, those individuals who exhibit a character that is adapted to the new conditions will have a better chance of surviving and of transmitting their peculiarity to future generations. In the contest of life 'a grain of sand turns the balance' (E, p. 115) and thus 'if a seed were produced with infinitesimal advantage it would have better chance of being propagated' (E, p. 137). Darwin now saw that some artificial breeds were in a sense adaptive, as with greyhounds which have been bred for speed in hunting. The natural process of selection is thus analogous to artificial selection but far more effective because it scrutinizes every minute character of every individual in every generation. Such a process would explain the branching adaptive evolution to which he was committed and would still allow him to believe that nature was designed by God to maintain species in a state of perfect adaptation to their environment.[28]

The role played by Malthus' principle may suggest that Darwin drew his inspiration in part from the individualist ideology of the middle classes.[29] Malthus' willingness to let the poor starve has often been seen as an early anticipation of 'social Darwinism' – yet it is clear that he did not see wealth as a prize to be gained in a free-for-all struggle between individuals of varying ability. Malthus presented his principle as God's method of encouraging us all to work hard (or face the threat of starvation) and this teleological viewpoint fitted in well with the young Darwin's thinking. He may also have been impressed by the quasi-numerical character of Malthus' demonstration, which fitted in with his search for testable laws. Yet in the end Darwin took from Malthus something that was not really there in the original: a sense that struggle could become a creative process by weeding out the unfit in every generation. It must be emphasized that Malthus only used the phrase 'struggle for existence' when discussing the competition between primitive tribes, not in the context of individualism. It is at least possible that this reference may have evoked Darwin's memories of the primitive natives being wiped out by European settlers. If we wish to confirm Darwin's commitment to the ideology of individualism, there are other examples besides Malthus on which we can base the case, since he was clearly well read in the works of Adam Smith and other political economists who had called for the establishment of a society based on free enterprise. We should also remember his family background among successful middle class industrialists and professionals.

Darwin's use of the struggle metaphor was soon transferred to the level of competition between the individuals making up a single population. This clearly links his thinking with the ideology of free-enterprise individualism. Evolution, like social progress, takes place because of the constant activities of the individuals who make up the population, as each struggles against its rivals to gain a living from the environment. Yet his theory cannot be taken as a typical expression of middle class values. Other thinkers, including Herbert Spencer, would use struggle as the basis for their biological and social evolutionism, but none saw the selective elimination of the unfit as a truly creative mechanism. Spencer depicted the struggle for existence as a spur to individual self-development and remained a lifelong Lamarckian. His system also included an element of inevitable progress that Darwin found impossible to square with his understanding of the irregular character of biological evolution. Malthus' emphasis on the pressure of population was capable of being exploited in many different ways. Darwin's use of the population principle was unique in his own time and contained the seeds of an idea that would outlast the optimistic progressionism of more typical Victorian evolutionists such as Spencer.

Darwin's continued belief that variation and struggle fulfil a divine purpose in keeping life adapted to an ever-changing earth suggests that he had not yet made a complete break with traditional beliefs. Even so, it is only in the vaguest sense that his theory can be seen as a modification of the orthodox Anglicanism of Paley.[30] Darwin was a deist rather than a theist: his mechanism had an ulterior purpose but it worked by the rigid application of natural law and left no room for the sense of a divine providence overlooking all the activities of living things.

The M and N Notebooks also make it clear that he had adopted an essentially materialist and determinist view of human nature. He accepted from the start that a theory of evolution must include mankind as a product of the animal kingdom. As early as the B Notebook he wrote, 'If all men were dead then monkeys make men. – Men make angels' (B, p. 169; note how easily Darwin himself falls into progressionism when discussing human origins). The M and N Notebooks contain numerous speculations about the process of human evolution. These notes anticipate most of the themes that would eventually emerge in the *Descent of Man* (see chapter 10 below). Darwin accepts that much of our unconscious behaviour is instinctive, programmed by evolution into the very structure of our brains. Even his private notes reveal how careful he saw he must be on such topics: 'To avoid stating how far, I believe, in Materialism, say only that emotions, instincts degrees of talent, which are hereditary are so because brain of child resemble, parent stock. – (& phrenologists state that brain alters)' (M, p. 57). He saw the various modes of expressing the emotions as a clear sign of our animal ancestry. He was now convinced that evolution would throw light on our moral values by showing how certain modes of behaviour would have become instinctive within any species that lived in social groups. Morality is merely the rationalization of these social instincts.

At a very early stage Darwin had thus faced up to the most radical implications of his evolutionism, freely accepting the consequences that would lead many of his contemporaries to resist the theory for several decades yet. It is clear that his wife Emma knew about his thoughts in this direction – indeed by the end of the N Notebook Emma had replaced his father as the authority to whom he looked for insights on human nature. Yet we know that Emma's more orthodox religious position must have led her to view with deep disfavour the more materialistic implications of the theory. Historians have argued endlessly over whether or not Darwin retained some form of religious belief during the decades in which he was developing his theory. Many now accept that, despite his recognition of the materialist implications of selectionism for human nature, he continued for some time to believe that the natural world was created by a rational God. It may be more fruitful, however, to follow David Kohn's suggestion that Darwin's thinking was actually enlivened by the tension between his materialistic and his theistic inclinations.[31] Instead of trying to thrash out a consistent position, he allowed himself to explore both directions as he switched his attention from one topic to another.

The same tension can be seen in Darwin's attitude towards evolutionary progress – and for the same reasons. There are many passages in the Notebooks which suggest that he still hoped to show that evolution was, in the end, progressive. At one time the majority of historians (myself included) tended to dismiss these remarks as lapses of concentration in which Darwin unwittingly fell back on ideas which he knew, in reality, had been undermined by his new theory. Many Darwin scholars now accept that his progressionist remarks should be taken more seriously and warn against the tendency to read the anti-progressionist viewpoint of modern Darwinism back into Darwin's own writings.[32] On this interpretation, the Notebooks reveal the thoughts of a man who was still enmeshed in the progressionist values of his culture. Darwin hoped to show that the values of European civilization were somehow nature's own values, in the sense that evolution was bound ultimately to generate higher organisms in which those values could be articulated. Far from wishing to destroy a sense of purpose in nature, he still expected evolutionism to support the belief that his own culture was the high point (so far) of the universe's activity. Mankind had to be treated as a product of material nature, but this view did not make nonsense of all moral values provided one assumed that nature was designed to enhance the development of those values through time.

At the same time we have seen that Darwin's sense of the branching nature of evolution made him very suspicious of conventional progressionism in which nature simply ascends a ladder of developmental stages towards modern humanity. The fact is that, as a biologist, he found himself questioning assumptions which – as a Victorian Englishman – he found it difficult to shake off. We should not try to pretend that Darwin fully appreciated all the anti-progressionist implications that modern biologists read into his theory. But nor should we try to dismiss him as just another Victorian progressionist depicting nature as an extension of middle class culture. Darwin is remembered today, when progressionists such as Herbert Spencer are forgotten, because he was led (perhaps unwillingly) to create a theory which had the potential to undermine the values of his time. If Darwin was a progressionist, he was a very strange one by Victorian standards. Where many others saw evolution as a necessarily progressive force, he realized that at best progress would follow only an overall statistical trend. Amidst the vast complexity of nature's efforts to adapt all her productions to their ever-changing environments, there would be a long-range tendency for some species to become more complex than any of their ancestors. If the human race was the high point of creation, it had reached its pinnacle via a very irregular course which could never have been predicted in advance.

Whatever the creative uses to which Darwin put his doubts about religion and the purpose of evolution, it seems clear that here was the source of the emotional tensions which may have begun to exaggerate his predisposition to stomach upsets and heart palpitations. He might be able to live with the implications of what he was doing but he knew that he was on very dangerous ground. Even his wife disapproved, and Emma's dismay both increased the psychological pressure and brought home to him the social dangers inherent in this course of research. If he stayed in London he would risk giving the game away to one or other of his more orthodox scientific contacts. Once it became known that he had adopted a theory with materialist implications, social ostracism might follow even if he himself still hoped that the system might be reconciled with belief in a remote kind of Creator. Various factors thus coincided to suggest that a move to a more remote location was desirable. From a country house near London Darwin could escape the pressures of the capital and yet stay in touch with those naturalists whom he might eventually hope to convert. The move to Down in 1842 was thus an almost inevitable consequence of the new turn in his scientific theorizing.

6

The Years of Development

THE MOVE TO Down House left Darwin free to develop
his ideas in relative security. The basic idea of natural
selection had already been clarified in his later notebooks and in 1842
he first wrote out a sketch of his theory. For the next seventeen years
he worked on his idea in relative isolation – yet it was an isolation that
could be relieved when necessary in accordance with his own wishes.
Far from seeing Darwin as a recluse, modern historians marvel at the
skill with which he built up a network of communication that would
serve his purposes. He corresponded widely with breeders and naturalists who could provide him with information that might bear upon
the thousand and one applications of a general theory of evolution. He
also built up a network of trusted contacts to whom he felt he could
safely expose his ideas. In this way he could elicit critical reactions that
would guide his own thoughts. He also hoped to create a small body
of influential scientists who might eventually help to introduce the
theory to a hostile world.

Darwin did not confine himself to gathering information from others,
however. Despite his illness he devoted himself to intensive study of a
number of areas of natural history that might throw light on the theory.
In particular, he produced an extensive description of the barnacles, a

hitherto largely ignored subclass. Darwin's monographs on the living and fossil barnacles helped to establish his credentials as a biologist but the work also allowed him to study the variability of populations and the implications of evolutionism for classification. In addition, he conducted practical studies of animal breeding and continued his work on biogeography

These studies did not merely provide additional evidence for the theory. At one time historians tended to assume that the whole evolutionary system was more or less complete by the early 1840s, explaining Darwin's failure to publish solely in terms of his unwillingness to face the social consequences of being branded a materialist. Darwin *did* fear the loss of respectability, but we now know that his refusal to publish stemmed at least in part from a recognition that the theory was not yet complete. Although committed to the metaphor of evolution as a branching tree, he did not yet understand why the branches of the tree constantly tended to diverge further and further away from one another. Only in the mid-1850s did he feel that he had a satisfactory explanation of this effect, and in the later years of that decade he began to write an extensive account of the theory for publication. This project was interrupted in 1858 by the arrival of Wallace's paper outlining his independently discovered theory of natural selection. Only then did Darwin begin to write the shortened account of his system that we know as the *Origin of Species*.

Man about Down

The Darwins moved into Down House in the village of Down (now renamed Downe), Kent, on the 14th of September 1842. Darwin's father paid £2,200 for the house, somewhat less than the asking price. In a letter to his sister Catherine, Darwin described it thus:

> House ugly, looks neither old nor new. – walls two feet thick – windows rather small – lower story rather low. – Capital study 18x18. Dining

room, 21x18. Drawing room can easily be added to is 21x15. Three stories, plenty of bedrooms.

The location was ideal for someone seeking solitude yet who wished to remain within easy reach of London:

> Position. – about 1/4 of a mile from small village of Down in Kent 16 miles from St. Pauls – eight miles & $\frac{1}{2}$ from Station (with many trains) which station is only 10 from London ... Village about 40 houses with old walnut trees in middle where stands an old flint Church & the lanes meet ... I never saw so many walks in any other country – The country is extraordinarily rural & quiet with narrow lanes & high hedges.[1]

Over the next few years the Darwins made a number of alterations to the house and gardens, including lowering the lane outside the front in order to increase their privacy. They soon had a comfortable family home in which Darwin felt that he would be happy to spend the rest of his life.

Down House can still be seen today in something like the state in which Darwin knew it. The Royal College of Surgeons maintains the building as a memorial to Darwin and it is open to the public. Several of the rooms have been restored, including Darwin's study. Visitors can also explore the extensive gardens including the 'Sand walk', a path around a small wood along which Darwin regularly walked for exercise.

Life in the new house began with a tragedy. Within three weeks Emma gave birth to a daughter, Mary Eleanor, who died within a few days. Over the next few years Emma bore a number of children: Henrietta in 1843, George in 1844, Elizabeth in 1847 and Francis in 1848. Charles and Emma were devoted to their children and visitors to the house often recalled the happy atmosphere. Darwin cared for his children as a father, but as an evolutionist he also observed their development closely in the hope of discovering clues to the origins of our mental functions and the ways in which we express emotions. As happened all too often in Victorian times, however, there was a price to be paid for this domestic happiness. In 1851 their first daughter, Anne, died at the age of ten.

PLATE 9 Down House

Poor dear little Annie, when going on very well at Malvern, was taken with a vomitting attack, which was at first thought of the smallest importance; but it rapidly assumed the form of a low and dreadful fever, which carried her off in ten days. Thank God she suffered hardly at all, and expired as tranquilly as a little angel. Our only consolation is that she passed a short, though joyous life. She was my favourite child; her cordiality, openness, buoyant joyousness and strong affections made her most loveable. Poor dear little soul. Well, it is all over. . . . [2]

There was nothing to do but comfort one another and carry on with the life of the family.

Darwin's routine at Down was designed to maximize the amount of work he could do in the face of his chronic illness.[3] He rose early and took a short walk before breakfast. His best work was done between 8.00 and 9.30 a.m., after which he read any incoming letters and had family letters or a novel read aloud to him. He worked again from 10.30 a.m. until noon, when he took another walk, often calling in at the greenhouse to check on his experimental plants. He usually circled the Sand walk several times, and this was also a favourite spot where the children played. After lunch he read the newspaper – he kept up an active interest in politics – and then wrote replies to his letters. In later years his son Francis often acted as his amanuensis. After a short rest, again accompanied by Emma reading aloud from a novel, he took another walk and then put in a final hour's work before dinner. He seldom lingered at the table after dinner since too active a conversation would often provoke a nervous attack that would spoil the next day's work altogether. He played two games of backgammon with Emma, read scientific books for a while and then listened to Emma playing the piano. He then retired but seldom slept well.

During the periods in which his illness flared up he could do no work and was completely dependent on Emma to nurse him. He had genuine fears that he would not live to complete his work and, as we shall see, took steps to ensure that Emma would arrange for publication of the theory in the event of his death. During a visit to his father at Shrewsbury in 1844 he wrote to Emma: 'I did not require to be reminded how well, my own dear wife, you have borne your dull life with your poor old sickly complaining husband. Your children will be a greater comfort to you than ever I can be.'[4] Emma ministered to him without complaint, her patience and cheerfulness serving as a foundation upon which the strength of the family could be built.

Robert Waring Darwin died on the 13th of November 1848. Darwin was so ill at the time that he arrived in Shrewsbury too late to attend his father's burial service. He inherited a fortune which has been estimated at £45,000, a considerable sum in those days. Over the years he increased this with a series of successful investments – for all his devotion to science he had a natural bourgeois talent for looking after money. The income from these investments was more than sufficient to maintain his growing family.

As the owners of one of the largest houses in the district, the Darwins would have been expected to play an important role in local society. James Moore has suggested that in fact Darwin relished the role of local 'squarson' (a combination of squire and parson).[5] Thanks in part to Emma's strong religious feelings, the family was closely involved with the local church. Even Charles got on well with the Reverend Brodie Innes, who became Vicar of Down in 1846, although the link with church affairs was severed in later years when Innes was replaced by a less sympathetic figure. Darwin helped to organize a Friendly Club, he was treasurer of the Coal and Clothing Club and he looked after the accounts of the National (i.e. Church of England) School and the Sunday School. From 1857 onwards he served as a county magistrate. In becoming a driving force in local affairs, Darwin seems to have satisfied a desire for respectability. The local people knew nothing of his unorthodox opinions and he was thus able to fit very nearly into the role that had originally been planned for him as a country vicar. By cutting off day-to-day contact with the scientific world of London, Darwin ensured that his potentially dangerous secret was safe for the time being.

In the *Autobiography* Darwin himself created the impression that his life at Down constituted an almost complete retirement from the world.[6] However, the isolation could be broken whenever it suited him. Darwin still needed contact with the world of science if he was to continue his work but the contact had to be on his own terms. In the early years he still went up to London fairly frequently, often travelling by the early train because this allowed him to spend the most active part of his day consulting with colleagues and looking for books and specimens. The penny post service, introduced in 1840, was very efficient, allowing him to interact with many individuals and institutions at arm's length. He joined a number of breeders' clubs and built up an unrivalled network of communication with people who could supply him with practical information on the subjects of variation and heredity. James Secord points out that many of the breeders were anxious to see their work put on a scientific basis.[7] They were thus happy to supply the information that Darwin requested in the hope that they would eventually be given credit in a responsible publication. Naturalists too were willing to correspond with someone who had already made his name through the publication of the *Beagle* voyage's results. Here again Darwin worked hard to set up a communications network that would allow him to test his ideas against the hard-won experience of others – without necessarily having to reveal his underlying purpose.

As he gradually began to sort out which naturalists were most likely to be sympathetic to his views, the contacts became more open and more direct. He made no secret of the fact that he was interested in the variability and stability of species – even the highly conservative Henslow knew this – and gradually a few selected contacts were informed of the full extent of his unorthodox thinking. We have already noted that by the early 1840s he was exploring the implications of evolution theory for biological classification with the zoologist George Waterhouse. By the middle of the decade at least half a dozen other scientists, including Lyell, knew something about the direction of his thoughts.

Perhaps the most important addition to this inner circle was the botanist Joseph Dalton Hooker, newly returned from his own voyage of discovery in the Antarctic. The son of Sir William Hooker, the Director of the Royal Botanic Gardens at Kew (a position he would eventually occupy himself), Hooker had met Darwin briefly in Trafalgar Square shortly before setting off for the Antarctic in HMS *Erebus*. On his return the acquaintance was renewed and within two months Darwin risked giving the whole game away:

> I was so struck with distribution of Galapagos organisms &c &c & with the character of the American fossil mammifers, &c &c that I

PLATE 10 *Joseph Dalton Hooker*, photograph reproduced by courtesy of the National Portrait Gallery, London

determined to collect blindly every sort of fact, which cd bear in any way on what are species. – I have read heaps of agricultural & horticultural books, & have never ceased collecting facts – At last gleams of light have come, & I am almost convinced (quite contrary to opinion I started with) that species are not (it is like confessing a murder) immutable. Heaven forfend me from Lamarck nonsense of a 'tendency to progression' 'adaptation from the slow willing of animals' &c, – but the conclusions I am led to are not widely different from his – though the means of change are wholly so – I think I have found out (here's presumption!) the simple way by which species become exquisitely adapted to various ends.[8]

Hooker was the first biologist to see Darwin's detailed exposition of his theory, the 'Essay' of 1844. Over the following years the two men conducted a lengthy debate on geographical distribution. In addition to his experience on the Antarctic voyage Hooker also spent the years 1847–50 botanizing in India. He was thus a mine of information on the geographical distribution of plants and an able critic of Darwin's views on the mechanisms of dispersal. Despite initial scepticism Hooker was, in the end, one of Darwin's first converts. Only in the 1850s did the much younger Thomas Henry Huxley attract Darwin's attention as another potential supporter who could be incorporated into the network.

Darwin chose his contacts well, since none of them reacted violently against his suggestions or denounced him publicly as a materialist. He was thus able to build up a small group of sympathizers who were prepared to talk seriously with him on the topic of transmutation and who were gradually won round to his way of thinking. Instead of assaulting conventional opinion head on, Darwin preserved his own respectability by seeking to undermine orthodoxy from within. He identified those figures who were least committed to old-fashioned creationism and gradually began to persuade them that a new approach to the species question was possible. When the time came to go public, his theory would be stronger for having withstood the private criticism of some of the best young naturalists in the country. This small but active and influential group would also form a fifth column already committed to the new idea and willing to promote it within the scientific community.

The interactions that sustained this network were not always limited to letters. Once Darwin was sure that he could trust someone, an invitation would be issued to visit Down for intimate discussions of the new approach. Such visits would often last for the whole weekend. In November 1845, for instance, Hooker and Waterhouse were at Down, along with Hugh Falconer and Edward Forbes. Francis Darwin later recalled the skill with which his father managed to encourage lively conversation among his visitors, especially when Huxley began to visit regularly.[9] Over the years Darwin thus managed to build up close personal relationships with the naturalists on whom he would depend for the eventual launching of his theory. These men became close family friends as well as scientific colleagues. Less intimate contacts and foreign visitors would often be entertained to lunch, easily possible as a day trip from London. Far from being a recluse at Down, Darwin had found himself the best possible position from which to conduct both his unorthodox researches and his campaign to subvert the scientific community.

Scientific Work

In 1842 Darwin wrote out a thirty-five-page sketch of his theory in pencil. In 1844 this was extended to a substantial 230-page 'Essay' which, at least in principle, was fit for publication.[10] This Essay is of considerable interest to historians because it gives a detailed outline of how the question of evolution stood in Darwin's mind at this early stage in his career. In many respects his exposition already looks forward to the *Origin of Species*. The basic sequence of topics has already been established: rather than beginning with the general case for evolutionism, Darwin launches straight into a description of his newly discovered mechanism. He first describes the effects of artificial selection and then goes on to outline how the struggle for existence will produce an equivalent form of selection by picking out those individuals best adapted to the environment. Later chapters attempt to meet some of the objections that Darwin foresees will be raised against his theory and then set out the advantages that naturalists will gain by switching from simple creationism to a theory of natural development.

The fact that Darwin chose to begin his argument with a description of the selection mechanism suggests that he already sensed the importance of a new initiative in this direction. He knew that at least some of the younger naturalists were becoming dissatisfied with simple creationism but were unwilling to make a move unless they could be convinced that something more promising than the now discredited Lamarckism was available. By starting his account in this way, Darwin ensured that a reader would be confronted immediately with the most original aspect of his theory. He thus hoped to overcome the kind of scepticism that arose not from a commitment to creationism but from mere frustration at the lack of any viable alternative.

Although the 1844 Essay anticipates much that would appear in the *Origin of Species*, historians now accept that in some respects it expresses only an immature version of the theory. Darwin had worked out the basic concept of divergent evolution due to natural selection operating upon isolated populations, but he had not yet explored all the implications of this idea and his views were to undergo significant development before the *Origin of Species* was written. As Dov Ospovat has pointed out, the original version of the theory was a compromise between static creationism and a totally dynamic model of natural change.[11] The chapter on natural selection makes it clear that Darwin still saw transmutation as an episodic rather than a truly all-pervasive process. A species only evolves when its environment is actually changing; once conditions become stable the species soon becomes perfectly adapted to the new state of affairs and there is no room for further modification. Since Darwin's pre-genetical view of reproduction was based on the assumption that variation is caused by a changed environment interfering with the process of individual growth, he believed that in a stable environment there would be little or no variation and hence nothing for selection to act on. Periods of evolutionary change would thus be interspersed with periods of stability in which a population of more or less identical organisms was perfectly adapted to the environment. The Creator had ensured that, when the environment began to change, the reproductive process would begin to generate variations on which selection could act to produce corresponding changes in the species. At this point in his career, Darwin had not realized that the pressure of population would maintain the struggle for existence even in a stable environment.

One consequence of this was that Darwin soon came to realize that his theory offered no explanation of an effect that palaeontologists were already demonstrating in the fossil record, the tendency for the lines of evolution to diverge ever further away from one another as the organisms specialize for different ways of life. Why did the evolution of the mammals, for instance, show a pattern of adaptive radiation in which numerous lines of specialization branch out to give modern forms such as the horse, the giraffe and so on? The 1844 Essay explains small-scale branching as in the Galapagos islands but not these major trends spread out over whole geological epochs. From the mid-1840s onwards Darwin began to feel that his theory was incomplete without a solution to this problem, and this is a point that needs to be taken into account when we ask why he refused to publish the 1844 Essay.

He certainly had not written the Essay for immediate publication. Its purpose was to ensure that his idea would not be forgotten if – as he genuinely feared – he should die prematurely. A letter dated 5 July 1844 instructs Emma that in the case of his sudden death she should use the sum of £400 to make sure that the Essay finds a publisher.[12] Several possible editors are suggested, including Lyell and Hooker. The fact that Darwin took this precaution suggests that he had no intention as yet of risking exposure as a materialist. 1844 was, in any case, the year in which Chambers' *Vestiges* aroused the anger of conservative scientists by suggesting that mankind was the end-product of a continuous progression in the animal kingdom. We know that Darwin thought very deeply about the impassioned arguments that Sedgwick raised against the book.[13] He had no time for Chambers' vague 'law of progress' and knew that his own theory offered a far more naturalistic explanation of organic development – yet by the same token it would come as an even greater shock to the exponents of natural theology. Darwin was determined to wait and try to build up support privately before risking publication, although it is significant that the revised edition of the *Journal of Researches* (1845) drops several broad hints that his observations have implications for the origin of species. Over the next decade there would be growing dissatisfaction with simple creationism among liberal members of the scientific community, creating a climate of opinion that would become more receptive to a new initiative on this topic. Yet it was only after he had solved the problem of divergence that Darwin decided to begin moving towards publication.

In the meantime Darwin continued his researches on a number of fronts. He bred pigeons and had their skeletons mounted to show the range of fundamental structural differences between the various breeds. He also continued to worry about the problems of geographical distribution. Here Hooker's wide experience with the plants of many different parts of the world turned out to be of major importance. Darwin could try out his views on the implications of evolution for distribution by measuring them against a fund of information even greater than his own. At the beginning of 1847 Hooker was given a copy of the Essay so that he could see the full extent of Darwin's theorizing. He was no easy convert, and the two men continued an intense dialogue over the next decade until Hooker eventually came to accept the new approach. One source of disagreement was Hooker's support for the view that the dispersal of species had often been aided by 'land bridges' thrown up across the modern oceans by geological movements in the recent past. Darwin was convinced that such large-scale elevations and subsidences of the earth's surface were geologically implausible and biologically unnecessary. He suspected that plant seeds and even the eggs of some animals could be transported over vast distances by wind, birds and ocean currents. This would explain why some forms of life seemed to reach oceanic islands much more readily than others and thus diversify into new species. To back up his views Darwin began an extensive study of dispersal mechanisms and performed experiments to show, for instance, that seeds could remain viable even after being immersed in sea water for considerable lengths of time.

By far the most extensive programme of research was that conducted into the barnacles. Darwin was led into this project through wishing to study an unusual species collected on the coast of Chile. He found that in order to understand the relationship between this specimen and the rest of the subclass he had to dissect and study members of other species. He soon began to realize that the whole subclass was so little known that it would be worthwhile undertaking a full-scale description and classification of all its members. The affinity of the barnacles to the crustaceans had only recently been recognized through a study of their larval stages; now Darwin was to set out on an eight-year project to uncover the complex relationships between the members of the group. So intense was his dedication that one of the Darwin children, on visiting a neighbour, asked 'Then where does he do his barnacles?',

PLATE 11 Charles Darwin in 1854 (by permission of the Syndics of
Cambridge University Library)

apparently under the impression that all grown-ups must be engaged
in this activity.[14]

Darwin's techniques are described in some detail by his son Francis.[15]
His dissecting table was a large board set in front of a window. He
preferred to do as much work as possible with the simple microscope
before turning to the compound, and used makeshift tools and appar-
atus to a surprisingly large extent. Despite this superficially amateurish
approach, he produced a major contribution to biology that could stand
independently of his species work. He also extended his range of
contacts within the scientific community as he appealed for specimens

of rare species and for information related to the project. The two volumes of his *Monograph on the Sub-Class Cirripedia* (1851–4) and the corresponding two volumes of the fossil barnacles are still regarded as the foundation stone for our modern understanding of the group.

There is some evidence that Darwin took up the barnacle project at least in part to establish his credentials as a biologist. In response to what he took as an implied criticism from Hooker, he wrote 'How painfully (to me) true is your remark that no one has hardly a right to examine the question of species who has not minutely described many.'[16] The barnacle monographs certainly ensured that when he published on evolution no such criticism could be levelled against him. In fact, Darwin learnt a great deal from his studies and there are many aspects of his research that – with hindsight – can be seen to throw light on his evolutionary thinking.[17] He was trying to understand how the various forms of barnacles could have been formed by a process such as natural selection in which every stage of development must have been of some use *at the time*. Natural selection cannot plan ahead; it can only favour those characters that confer an advantage in the current situation, and this imposes limitations that are difficult for anyone wedded to a more teleological view of evolution to appreciate. As he described the various barnacle species Darwin was trying to work out how such bizarre forms could have originated. The adult stages of all barnacles are degenerate, and some have minute males that live almost as parasites within the females. Darwin had to see whether there was any advantage for them in adopting such an unusual reproductive strategy. Such problems would only have made sense to an evolutionist who was firmly convinced that something more complex than a simple progressive development was involved.

Darwin also began to realize how extensive was the amount of variation available for natural selection to work on. It may have been partly this consequence of the barnacle work that brought home to him the weakness of his original assumption that there is little variation in a population exposed to stable conditions. This in turn highlighted the problem of divergence: why had natural selection acted to produce such a high level of diversity within many groups? By 1847 it is clear that Darwin recognized the deficiency of the 1844 Essay on this question, a recognition that would be reinforced by the palaeontological work of Owen and others over the next few years.

The solution came to him only in November of 1854 when he first conceived the outline of his 'principle of divergence'. He now realized that natural selection was not the only factor that must be taken into account in explaining divergent evolution. The ecological pressures acting on the population were also important, and Darwin now became convinced that speciation (that is, the splitting up of one species into many) and divergence would take place most actively in locations where there was a high level of competition for resources. In January 1855 he realized that his idea was analogous to the concept of the 'division of labour' employed by economists (yet another link between his thinking and the social environment of the time). As Adam Smith had shown, labour is more productive if it is specialized so that each worker concentrates on a single aspect of the job instead of trying to make the complete article himself. Darwin saw that a given area of land would support a greater volume of life if the inhabitants were specialized to exploit the available resources in a variety of ways. Here was a form of ecological pressure that would allow natural selection to favour increasing levels of diversity even in a perfectly stable physical environment.

Over the next few years Darwin tested this idea by undertaking arithmetical studies of the number of species within different genera. As David Kohn has shown, this study was based on a form of historical reasoning which is characteristic of Darwin's whole approach to evolution and which sets it apart from the thinking of the vast majority of his contemporaries.[18] He tried to use different modern genera as models for the successive stages in the historical process by which a genus expands and contracts in the course of its existence. A successful new form begins to expand into its neighbouring territory and soon divides to give a small group of closely related species. As the process continues, it expands over more territory and divides into even more species. Almost inevitably, however, it will sooner or later be overtaken by rivals who begin to exterminate the component species. The final stage, just before the extinction of the whole group, will consist of a few widely scattered and highly divergent species. Darwin came to believe that he could provide examples of genera at all these various stages of development, confirming that competition and divergence were ever-active processes in the natural world. At a time when most other naturalists saw the goal of evolutionism as the construction of family trees linking fossil and living forms, Darwin was extending the view – already pioneered in his Galapagos studies – that the best illustrations of what is really happening in nature are to be found in the relationships between species and their environments in the world today.

The end-product of all this effort was a theory of evolution that differed significantly from that outlined in 1844. Darwin had now been forced to recognize that competition was an ever-present force in nature that would permit no episodes of stability in which every individual was perfectly adapted. There is always some variation, and the pressure to specialize will allow natural selection to change species even if their environment remains the same. One consequence of this was that he now began to argue that geographical isolation (as in the Galapagos) was not the only, nor even the best, situation in which speciation could take place.[19] His new emphasis on the power of ecological specialization led him to suppose that the best conditions for speciation would be a continuous area where there was massive competition for resources. Here the pressure to specialize would be so great that it could separate populations into non-interbreeding groups even if there were no geographical barrier between them. This was a position that would cause him considerable difficulty in later years, and many modern biologists believe that he was wrong to accept what is now known as 'sympatric speciation'. The problem is that, without a geographical barrier, incipient changes tend to be wiped out by interbreeding between the potentially divergent forms.

A far more pervasive change was also taking place in Darwin's religious views, although the full implications of his new approach would only become clear to him in later years. Even in the late 1830s he had abandoned any notion of a God who carefully superintends the process of evolution. He was not yet an atheist, though, since he still regarded evolution by natural selection as the kind of process that might have been instituted by a benevolent God to ensure that His creatures can maintain themselves in a world subject to geological change. His concept of 'perfect adaptation' would certainly have encouraged this view. But in the new model of evolution there was never any respite from the struggle for existence or the pressure of ecological competition. At first Darwin did not realize that this might threaten his belief that the natural system was established by a benevolent God. The *Autobiography* records that he was still a theist when he wrote the *Origin of Species*.[20] A theist believes in a God who not only created but also cares for and sustains the world – as opposed to a deist who believes merely in a remote and uncaring Creator. Darwin thus tried to sustain his original faith in a God who minimized the amount of suffering in the world, but as the years went by he found it increasingly difficult to believe that the relentless struggle for existence was a product of divine contrivance. He was still unwilling to abandon the possibility that in the long run the process of adaptation might be seen to have an overall purpose. Yet throughout the rest of his career he would find himself plagued with doubts about the possibility of any meaningful relationship between the Creator and the individual events that take place in the world of living things.

Apart from his own slowly ebbing preference for a theistic view of nature, Darwin realized that he must be very careful to minimize the materialistic aspects of his theory in any public pronouncement. He knew that, whatever the growing dissatisfaction with creationism, the vast majority of naturalists and ordinary people would only be willing to tolerate a process of 'creation by law' if they felt that the law somehow expressed a divine purpose. In the 1844 Essay he had even introduced the concept of natural selection by first creating the image of a quasi-divine overseeing Power which could pick out useful variants just as the animal breeder does in a domesticated species.[21] It is obvious from later sections of the Essay that there is no such superintending Power, since natural selection works without forethought and depends solely on the day-to-day operations of the most ordinary natural laws. The metaphor of the superintending Being was merely a device that would help those with theistic beliefs to come to grips with the idea. The metaphor does not appear in the *Origin of Species* – yet, as Robert Young has pointed out, the very term 'natural *selection*' helped to encourage the view that nature was, after all, an intelligent agent.[22] It was in Darwin's own interest to preserve as much as possible of the traditional view that natural development represented the unfolding of a divine purpose.

Darwin could no longer be a simple-minded progressionist, given that his theory of divergent evolution predicted no single goal towards which all living things were advancing. The difficulty of comparing forms with entirely different fundamental structures led him to wonder whether the very concept of one organism being 'higher' than another could be translated into meaningful biological terms. In his copy of Chambers' *Vestiges* he pencilled a note: 'Never use word higher and lower'.[23] It is thus possible to argue that at least part of the time he was aware of the anti-progressionist implications that many modern biologists read into his theory. The theory of divergent evolution had the potential to undermine any concept of the inevitability of progress and could thus be difficult to reconcile with the view that nature was intended to develop towards a certain goal by its Creator. The crucial question is: did Darwin himself – and his followers after 1859 – fully appreciate this aspect of the theory?

In fact Darwin always found it difficult to stand by the most radical implications of divergent evolutionism. Even in his private letters to Hooker he seems to have retained the view that natural selection must, in the long run, generate higher levels of organization. Long exposure to competition would make modern species more efficient than their ancestors in coping with the environment and this 'will ultimately make the organization higher in every sense of the word'.[24] Many historians interpret this as a residual effect of his original belief that nature is a divinely created system.[25] The Creator had ensured that life would slowly but inevitably ascend to higher levels of development, leading to the eventual emergence of thinking beings such as mankind. When Darwin at last came to write the *Origin of Species* he did everything in his power to create the impression that natural evolution worked for the long-term benefit of living things. Whatever his own reservations about the Creator's benevolence, he knew that his new approach would have a chance to influence the views of others only if it could be presented in a form that would minimize the widespread fear that evolution was a step towards atheism.

By 1856 Lyell and several other friends were encouraging Darwin to publish his idea, warning that if he delayed too long he might be forestalled by someone else. In May of that year he began work on a 'big book' to be entitled 'Natural Selection' and the writing went ahead steadily over the next couple of years. By the spring of 1858 he had completed ten chapters, covering about two-thirds of the material that would eventually be included in the *Origin of Species*. There seems no doubt that he was now convinced that the time was ripe for yet another attempt to float the idea of evolution before the public. He had already aroused interest in his idea among a significant number of naturalists and believed that publication of his theory would force even more to rethink their attitude. Robert Stauffer, the editor of the modern edition of the 'Natural Selection' manuscript, estimates that had Darwin completed it he would have published a substantial two-volumed work sometime in the early 1860s.[26] It is a matter of conjecture what effect such a long description of the theory would have had. On 18 June 1858 Darwin received the manuscript of a paper by Alfred Russel Wallace outlining an independently conceived theory of natural selection. This event was to change his plans completely and lead to the writing of the *Origin of Species* as we know it today.

7

Going Public

T HE CIRCUMSTANCES UNDER which the *Origin of Species* was written were unusual enough to have generated a great deal of speculation. The fact that two naturalists independently discovered the principle of natural selection has aroused the interest of those who like to claim that the theory reflects the ideology of Victorian Britain. Wallace is sometimes treated as a small-scale version of Darwin himself (*Darwin's Moon* is the title of one biography) while other historians insist that he was treated badly by Darwin and his supporters. Some even claim that Darwin stole the principle of divergence from Wallace. The desire to rescue Wallace from oblivion is laudable enough, but its more deliberately anti-Darwinian manifestations do not stand up to close scrutiny. Darwin conceived his theory of natural selection twenty years before Wallace and it is impossible to imagine Wallace being in a position to write anything on the scale of the *Origin of Species* in 1858. Closer study of Wallace's work reveals that it is in any case a mistake to treat him as merely a runner-up after Darwin. The two naturalists approached the question from different angles and differed on important points throughout their careers. A biography of Darwin is not the place for a detailed study of Wallace's work, but we should at least credit him with playing an independent role in the drama.

There remains the *Origin of Species*, Darwin's best-known and most frequently reprinted book. It is in some respects an unlikely detonator for the cultural explosion that it is supposed to have set off and some modern readers may wonder what all the fuss is about. Although Darwin wrote it as a relatively short introduction to his theory, it still comprises nearly 500 pages of detailed natural history. As scientific books go it is reasonably accessible to the non-specialist, but one sometimes wonders how many modern readers are able to follow the whole logic of the argument that Darwin develops amidst all the details. There are few discussions of the broader issues raised by the theory, although Darwin's opposition to simple creationism is obvious enough. To clarify the nature of the theory set out in the book the second section of this chapter will be devoted to an outline of its argument, in the hope that this will guide modern readers to Darwin's own expression of his ideas.

Enter Wallace

Wallace came from a poor background, lacking the many advantages of the Darwin family. He began his career as a land surveyor and then became a schoolteacher in Leicester.[1] Here he met an amateur entomologist, Henry Walter Bates, who encouraged his interest in natural history. In 1848 the two men set off for South America to collect zoological and botanical specimens, having been assured that there was a ready market for these in London. Wallace also intended to gather information on the geographical distribution of species with a view to studying their origin. He had been impressed by Chambers' *Vestiges* (far more so than Darwin) and was also a convert to Lyell's uniformitarian geology. On his return to England in 1852 Wallace's whole collection was destroyed by fire but he was insured and the disaster was thus scientific rather than financial. In 1854 he set off on another collecting expedition, this time to the islands of the Malay archipelago, modern Indonesia.

PLATE 12 *Alfred Russel Wallace*, photograph reproduced by courtesy of the National Portrait Gallery, London

In 1855 Wallace published his first theoretical paper in which he argued that a new species always comes into existence in an area already occupied by a related species.[2] The evolutionary implications of this claim are obvious enough, although Wallace offered no explanation of how the new species is actually formed. Darwin read the paper and was interested to note that it used the metaphor of a branching tree of relationships, but he saw no evidence that Wallace had anything to offer on the question of an evolutionary mechanism. The paper did have an impact on Charles Lyell's thinking on the species question, however, and may have been responsible for Lyell's recommendation that Darwin should write up his theory for publication. Meanwhile Wallace continued to think about the question, and in 1858 the solution came to him while he was suffering from a bout of fever on the island of Gilolo (not, as he later claimed, on the better-known spice island of Ternate). Already familiar with Malthus' views on population, he realized that, if a species existed in a number of variant forms, those that were poorly adapted to any change in the environment would be exterminated, leaving only the most well-adapted form. This was the idea that he wrote up in a short paper and sent off to the one man whom he thought most likely to take an interest and encourage its publication: Darwin.

When Darwin received the paper on 12 June 1858 he began to panic. Here was another naturalist who seemed to have anticipated the most important aspect of his own theory. He immediately consulted with Lyell and Hooker, who advised him to arrange for the simultaneous publication of Wallace's paper and (to ensure priority) a short extract of his own. The joint papers were read to the Linnean Society and then published in the Society's *Proceedings* under the title 'On the Tendency of Species to Form Varieties; and on the Perpetuation of Varieties and Species by Natural Selection'. They consist of (a) a short extract from Darwin's manuscript, (b) part of a letter that Darwin had sent to the American botanist Asa Gray in 1857 (this demonstrated Darwin's priority) and (c) Wallace's paper 'On the Tendency of Varieties to Depart Indefinitely from the Original Type'.[3]

It is sometimes argued that Darwin should not have subordinated Wallace's paper to his own in this way. Between them, Darwin, Lyell and Hooker ensured that Wallace would appear as the junior author, even though he had been the first to come forward with a publishable paper. Against this must be set the fact that Darwin had been working on his theory for twenty years and clearly had a much more extensive grasp of all the implications. A more serious charge raised by some historians is that Darwin actually plagiarized his principle of divergence from Wallace.[4] According to this interpretation Darwin concealed the fact that Wallace's paper had actually arrived some time before 12 June and that in the meantime he had used Wallace's work to complete his own theory of divergence and add it to chapter 4 of the 'Natural Selection' manuscript. There is no doubt that a substantial addition to the manuscript was made at this time, but the historians who accuse

Darwin of plagiarism ignore the fact that – as the letter to Gray in 1857 was intended to demonstrate – Darwin had already outlined his interpretation of divergence long before Wallace even wrote his paper.

The efforts to denigrate Darwin serve only to conceal the real differences between the two naturalists' approach to transmutation. Careful reading of Wallace's paper reveals that in several important respects his theory failed to duplicate the essence of Darwin's thinking. Wallace had no interest in artificial selection and refused to treat it as analogous to the natural process even in later years. His mechanism did not even address the basic question of how selection acts on individual differences to change a population, because he was interested in how one well-marked variety (what we now call a subspecies) could replace others. Once it is recognized that in writing of natural selection acting on *varieties* Wallace was thinking of subspecies rather than individual variations, it can be seen that his paper does not contain a description of what Darwin saw as the basic mechanism of change. Wallace simply assumed that species split into varieties – he did not seek to explain how this all-important first step occurs. It has also been suggested that Wallace failed to appreciate the full power of selection because he treated the varieties as struggling against nature, not against each other.[5]

Darwin may thus have overreacted on receiving Wallace's paper; his fear that he had been anticipated arose because he read more into the paper than was really there. Wallace never admitted that his 1858 paper was not a complete expression of the Darwinian theory and in later reprintings he added subtitles that help to blur the distinctions. But it is significant that – unlike his modern defenders – he never complained about the way in which the publication of his paper had been handled. Wallace went on to make major contributions to the study of geographical distribution and to engage in a fruitful dialogue with Darwin on how natural selection works. Far from being rivals, the two men always treated each other with respect and, later on, with affection.

The Linnean Society publication went virtually unnoticed. Even the Society's president complained at the annual general meeting that the year had not been marked by any striking discoveries. This suggests that short papers were not enough to arouse interest in so momentous a subject. But Darwin had now been goaded into action, determining to write a single-volume account of his theory to put before the public

as quickly as possible. He began work while on holiday on the Isle of
Wight in July and afterwards complained that the writing had been
restricted by constant bouts of pain. John Murray agreed to publish
the book but received conflicting advice on how well it might sell and
in the end printed only 1,250 copies of the first edition. Darwin was
persuaded not to call it an 'abstract' of a larger work, finally settling
on the title with which we are now familiar: *On the Origin of Species by
Means of Natural Selection: or the Preservation of Favoured Races in the
Struggle for Life*. He began proof-reading on 25 May 1859, making
significant corrections as he went, and finished on 10 October. The
book came out on 24 November at a price of 15 shillings and was sold
out to the bookshops on the first day. Darwin retired to Ilkley to take
the waters in anticipation of the coming storm.

The Argument of the Origin of Species

The *Origin of Species* is Darwin's best-known book and is widely available
in modern reprints. It should be noted, however, that in response to
numerous criticisms Darwin undertook constant revisions between the
book's first appearance in 1859 and the sixth edition of 1872. The later
editions thus differ considerably from the first, and the last edition contains
an additional chapter (chapter 7) dealing with objections to the theory.[6]
These changes tend to obscure the original argument and the first
edition is thus by far the clearest expression of Darwin's insight. A
facsimile of this edition is available and also a concordance, although
many modern reprints unfortunately follow the text of the sixth edition.
All references below are to the first edition unless otherwise stated.

The 'Introduction' leads the reader directly to the question of adaptation which Darwin sees as the central problem of evolution. He argues
briefly that Lamarckism cannot explain all cases of adaptation and
points out that the simple progressionism of the *Vestiges* does not
even address the question of how species become adapted to their
environment. From the start Darwin makes it clear that he will be
offering a new mechanism to explain the origin of species by adaptation.

The main text of the *Origin of Species* can be divided into three parts. Chapters 1–5 outline the theory of natural selection, chapters 6–9 (6–10 in the sixth edition) deal with the many objections that Darwin foresees will be raised against his theory, and the concluding chapters (10–14 in the first edition, 11–15 in the sixth) show how a wide range of otherwise inexplicable phenomena can be illuminated by the theory of common descent by adaptive modification.

Chapter 1, 'Variation under Domestication', begins by stressing that animal and plant breeders have been able to produce immense changes within domesticated species. Darwin was convinced that the analogy with artificial selection was the best way of helping his readers to understand how nature might be able to produce similar changes by an equivalent selective process. Almost immediately we encounter Darwin's pre-Mendelian theory of variation and inheritance. Like the notebooks of the 1830s, the *Origin of Species* is pervaded by a way of thinking about reproduction which is quite alien to modern biology. There is no direct equivalent of the modern belief that the population contains a reservoir of genetic variability, each gene being transmitted as a unit from one generation to the next. Instead, Darwin argues that individual variation is due to the direct effect of changed conditions on the reproductive process. This, he believes, will explain why domesticated species exhibit more variability than wild ones. Further details are given in chapter 5, 'Laws of Variation', where he insists that all changes of structure are *caused*, even though we do not know the cause and speak of them as being due to chance (pp. 131–2). In this chapter Darwin admits a minor role for Lamarckism, although he insists that most of the variability is undirected.

In his discussion of variation under domestication, Darwin moves from a general account of the 'random' variation in domesticated species to the work of the animal breeders who exploit that variation to produce significant changes. He goes to some length to show that we cannot draw a rigid distinction between the production of new *breeds* by man and the appearance of new *species* in nature. Artificial breeds do not invariably tend to revert to some fixed 'natural' form for their species. The diversity of pigeon breeds is enormous, and Darwin argues that, if the different breeds were shown to an ornithologist who did not know that they were domesticated, he would classify them as belonging to distinct species, perhaps even distinct genera. What, then, is the explanation of man's ability to produce such enormous changes in a species? 'The key is man's power of accumulative selection: nature gives incessant variations; man adds them up in certain directions useful to him' (p. 30).

In chapter 2, 'Variation under Nature', Darwin asks whether the variation which serves as the raw material of selection exists in wild populations. He admits that there may be much less variation in the wild because – according to his theory – wild species exist under 'natural' conditions which do not disturb the reproductive process. But individual differences do occur and even the most important structures are subject to variation, as evidence of which Darwin cites studies by John Lubbock which had demonstrated major variations in the patterns of the nervous system within a single insect species. 'These individual differences are highly important for us, as they afford materials for natural selection to accumulate, in the same manner as man can accumulate in any given direction individual differences in his domesticated productions' (p. 45).

Much of chapter 2 is devoted to a very different method of demonstrating the variability of wild species. Darwin points to the fact that many species are known to form fairly distinct and permanent races or 'varieties' adapted to the local conditions in certain parts of their territory. Naturalists had been quite willing to assume that such varieties were formed by common descent from a single ancestral form, yet there was often disagreement as to whether a particular form was a variety or a distinct species. Darwin argues that the confusion arises because varieties are merely an intermediate step in the production of new species: 'a well-marked variety may justly be called an incipient species' (p. 52). The term 'species' is thus an arbitrary one: a species is just a strongly marked variety that the majority of experienced naturalists agree to call a species. Once this point is accepted the barriers supposedly separating species have been broken down and the way is cleared for transmutation to be used to explain the production of all new forms of life.

Chapter 3 introduces the 'Struggle for Existence' which arises from the tendency of all species to over-reproduce. If more are born than can possibly survive, there must be competition to see which individuals will get enough of the scarce resources to keep themselves alive. Darwin is quite clear that this idea has its origins in the 'principle of population' that Malthus applied to human society: 'It is the doctrine of Malthus applied with manifold force to the whole animal and vegetable kingdoms; for in this case there can be no artificial increase of food, and no prudential restraint from marriage' (p. 63). The force of this argument is based on numerical calculations designed to show the potential rate of population increase that must be checked by the shortage of resources.

The struggle is most severe between members of the same species, or between closely related varieties, because here the individuals are competing for exactly the same resources. Many different factors determine who shall live and who shall die. Often the elimination takes place among the young; sometimes shortage of food is critical but often it is predators which keep the population down. Darwin emphasizes the complex web of interactions between species, each being held in check by others which prey upon it or upon which it preys. Many readers of the *Origin of Species* were impressed by his emphasis on the complex network of living relationships even if they did not accept the theory of natural selection. Darwin admits that behind 'the face of nature bright with gladness' there is a scene of constant struggle. Yet he was anxious not to create too harsh an image of nature. He wished to imply that his theory modified, but did not challenge, the prevailing belief that nature had been created by a wise and benevolent God. The last thing he wanted was for people to see his book as the basis for an amoral view of nature. He thus ends his chapter on the struggle for existence with the following words:

> When we reflect on this struggle, we may console ourselves with the full belief, that the war of nature is not incessant, that no fear is felt, that death is generally prompt, and that the vigorous, the healthy, and the happy survive and multiply. (p. 79)

Chapter 4, 'Natural Selection', contains Darwin's main description of his mechanism. After reminding his readers of the power of artificial selection and of the dependence of the wild organism on its environment, he goes on:

Can it, then, be thought improbable, seeing that variations useful to man have undoubtedly occurred, that other variations useful in some way to each being in the great and complex battle of life, should sometimes occur in the course of thousands of generations? If such do occur, can we doubt (remembering that many more individuals are born than can possibly survive) that individuals having any advantage, however slight, over others, would have the best chance of surviving and of procreating their kind? On the other hand, we may feel sure that any variation in the least degree injurious would be rigidly destroyed. This preservation of favourable variations and the rejection of injurious variations, I call Natural Selection. (pp. 80–1)

Darwin contrasts nature's powers with those of mankind, stressing how much more effective will be the scrutiny of nature. He points out that natural selection acts only to improve the organisms' ability to cope with their environment and thus tries to persuade his readers that, whatever the apparent harshness of natural selection, we can nevertheless see it as a force promoting the improvement of living things.

Reproduction rather than survival is the crucial factor of course, a point illustrated in the section on 'Sexual selection' in which Darwin argues that any character useful in obtaining a mate will become highly developed. The antlers of deer and the bright colours of many birds are explained by the fact that the mating habits of the various species allow those males with well-developed secondary sexual characters to attract more females.

Darwin explains that the extinction of some forms will be inevitable in a world governed by natural selection. Species will often be exterminated by competitors, thus in effect leaving room for the more successful species to multiply. At the end of chapter 4 he includes a long discussion of the process of divergence by which a single original form can give rise to a family of descendants, each of which will become more specialized for its own way of life and which may itself subdivide. Natural selection will always favour increased specialization even in a stable environment. He gives a diagram to illustrate the process – a classic example of an evolutionary tree. Significantly, Darwin's tree has no central trunk or 'main stem' of evolution. No one branch can be singled out as the main line of development because each is adapting in its own way to the environmental changes it encounters in the course of its migrations. There is no way in which the human race can be seen as the goal towards which the evolution of the whole animal kingdom has been striving.

With chapter 6, 'Difficulties on Theory', we move into the central section of the *Origin of Species* in which Darwin attempts to defend his idea against the objections he foresees will be raised against it. (The additional chapter added to the last edition, 'Miscellaneous Objections to the Theory of Natural Selection', includes Darwin's response to critics who had indeed expressed and extended the problems he had anticipated.) The first problem in chapter 6 is that of the lack of transitional forms between known species. This is an important point because it is often assumed that a theory of continuous evolution must imply that there are no gaps between species. Darwin argues that this is not the case if evolution is a branching divergent process. Divergence takes place through the continued extermination of the less specialized forms and thus the intermediates do not survive through to the present. In his view, 'species come to be tolerably well-defined objects, and do not at any one period present an inextricable chaos of varying and intermediate links' (p. 177).

The next problem is the origin of species with peculiar habits or structures. How, for instance, could a non-flying mammal evolve into a bat – surely the intermediate forms would have limbs that were not well adapted either for walking or for flying? Here Darwin appeals to the existence of flying squirrels with varying abilities to glide from tree to tree. These confirm that the intermediate state between legs and wings is viable, thus illustrating the route selection may have taken in the creation of the more perfect wings of bats. Darwin notes that species do indeed change their habits, as witnessed by certain kinds of woodpeckers that do not live in trees and web-footed geese that do not live in water. These examples are inexplicable on the assumption that God creates every species perfectly adapted to its way of life, but are to be expected if species are constantly trying to find new ecological niches to exploit and if evolution takes some time to adjust a species to a new way of life.

Darwin also refers to the problem of explaining the evolution of a highly complex structure such as the human eye. He notes that in the invertebrates there are creatures with eyes of varying degrees of complexity, showing that intermediate stages in the development of vision can be of benefit to the species. He admits that, if one could find a complex organ for which no intermediate states were conceivable, his theory would have to be rejected, but insists that no such organs are known. Another problem is the existence of organs of little apparent importance. If natural selection works only by seizing upon advantageous variations, then surely all characters must be adaptive. Yet many naturalists were convinced that species possess useless characters. Darwin responds to this problem by suggesting that we may often underestimate a structure's usefulness. The giraffe's tail seems to be used as a fly-flapper, which appears rather trivial until one recollects that in some parts of the tropics the distribution of large mammals is critically determined by their vulnerability to insect pests.

Chapter 7 deals with 'Instinct', a topic of particular interest to Darwin because he was convinced that evolution must be able to explain animal behaviour. The Lamarckians had an obvious explanation of instinct as a learned habit that had gradually been built into the species' hereditary constitution. But Darwin notes that this will not explain the instincts of neuter insects. He was convinced that natural selection can act on instincts just as it can act on physical characters. There is variation within the instinctive behaviour of any particular species, as shown by the fact that man has been able to eradicate the dog's instinctive tendency to attack sheep etc. Thus selection will be able to enhance a useful instinct by seizing upon useful variations. The case of neuter insects is explained by the fact that selection can act on families as well as individuals: insects with a tendency to produce some useful but neuter types among their offspring would survive better as the founders of colonies.

In chapter 8, 'Hybridism', Darwin deals with what many critics saw as a fatal objection to his theory. Varieties formed within the same species can be cross-bred with one another, but on the traditional view any attempt to hybridize distinct species will always fail. Darwin counters this argument by showing that the supposed absolute distinction between varieties and species is not as clear-cut as was popularly supposed. He cites extensive evidence to show that varying degrees of sterility are encountered in efforts to cross-breed different species of plants. If the species are closely related, there will often be a small degree of fertility in the hybrid offspring. His conclusion is that, far from being a fatal objection to his theory, a close study of hybridization confirms that there is no sharp distinction between varieties and species. As two related forms diverge further from one another in the course of evolution, their ability to interbreed gradually diminishes and eventually falls to zero, that is, absolute sterility.

Chapter 9 is 'On the Imperfection of the Geological Record'. Darwin had committed himself to the belief that evolution always takes place slowly and gradually, but he was aware that this did not seem to be supported by the fossil record which generally shows new species appearing abruptly with no sign of an evolutionary ancestry. He argues that the discontinuity is a result of the record's imperfection and not an indication that species really are introduced suddenly. He points out (p. 280) that it would be a mistake to look for simple intermediates between two related forms; they will have diverged from a common ancestor that would not have been an exact intermediate between its later descendants. But even this more complex kind of relationship is seldom found in the record, and Darwin insists that we should not expect to find all the steps in evolution preserved in the rocks. Fossil-bearing rocks are only laid down in certain circumstances, and thus long periods of time will have passed between the deposition of strata that now seem consecutive. Even when we have a continuous sequence of depositions, evolutionary changes may often have taken place in isolated parts of a species' range where fossils are not being laid down (a point amplified in the modern theory of punctuated equilibrium).

A related problem concerns the abrupt appearance of whole groups of living things at certain points in the record. The most obvious example of this is what we now know as the 'Cambrian explosion' – the sudden appearance of all the basic modern types at the beginning of the Cambrian era. (Note that in the first edition Darwin places this in the Silurian – the geological nomenclature was disputed at the time.) Darwin argues that the imperfection of the record also accounts for these episodes. There must have been vast periods of time before the Cambrian when life was evolving but from which so far no fossils had been found. Perhaps the continents were positioned differently so long ago, and most of the fossil-bearing rocks from that age are sunk beneath the modern oceans. All that Darwin could do was to express the hope that some fossils from the pre-Cambrian rocks would eventually be found to fill in the gap.

The next chapter continues the discussion of the fossil record but Darwin now moves onto the offensive so that his concluding chapters can present the positive case for evolution. 'On the Geological Succession of Organic Beings' argues that, if we make allowances for the imperfection of the record, the known fossils are distributed just as one would expect on the basis of a theory of common descent. Palaeontologists were generally agreed that in any sequence of fossil formations the intermediate forms in time were intermediate in character. Owen and others had also shown that ancient types often seemed to 'fall between' the more distinct modern forms. Thus the pigs and camels could now be united into a single group by including fossils which were intermediate in character. To Darwin it was obvious that these ancient types were the generalized ancestors from which the more specialized modern forms had evolved.

The next two chapters deal with 'Geographical Distribution' and include some of the evidence that converted Darwin himself to a belief in evolution. Darwin insists that the differences between the Old and New World faunas cannot be explained in terms of climate since both areas share the same range of physical conditions. The ostrich of Africa and the rhea of South America are superficially similar, but when examined carefully each is seen to be characteristic of its own continent. The chief determinant of geographical distribution is the barriers which exist to free migration, of which the oceans are the most important in the case of land animals. Conversely, dry land creates an obvious barrier to the migration of marine forms. Darwin argues that a successful species will spread out as far as it can, adapting to the local conditions it encounters, until it meets an impassable barrier. Thus on his theory it is to be expected that major barriers to migration will define unique groups of species. He also discusses the process by which oceanic islands such as the Galapagos are populated, bringing in his own studies of how seeds and eggs can occasionally be transported across wide stretches of ocean.

The penultimate chapter deals with 'Mutual Affinities of Organic

Beings' and allows Darwin to develop the explanatory power of his theory to the full. He picks out a number of phenomena encountered by naturalists in their efforts to classify species and demonstrates that they are explicable only in terms of a theory of common descent. The basic system of classification entails the grouping of similar species into genera, the genera themselves into families and so on. By uncovering these relationships naturalists were seeking what was called a 'Natural System' of classification – but Darwin now asks what this system is supposed to be. Is it the plan of the Creator, as many exponents of natural theology liked to claim? Darwin insists that to assume the existence of an underlying divine plan adds nothing to our knowledge of natural relationships. His theory alone can explain why species are grouped together: the natural system is an expression of evolutionary relationships, in effect a cross-section of the evolutionary tree.

Darwin next turns to embryology, noting that the embryos of different animals often show a much greater degree of resemblance than the adults. This, he believes, can be explained on the assumption that adaptive modifications are produced mostly by changes in the later stages of growth, leaving the early pattern of development unchanged. For Darwin, the embryo represents the species in a less modified state and thus helps to reveal its natural relationships. In some cases the embryo may actually resemble ancestral forms that can be discovered in the fossil record (pp. 449–50). But this is a far cry from the so-called 'recapitulation theory' in which the development of the human embryo is supposed to repeat the whole pattern of evolution revealed by the fossil record. Darwin had no interest in the idea of a linear pattern of development leading up to mankind as the goal of creation, and hence no interest in using the growth of the human embryo as a model for the 'main line' of evolution.

Finally, Darwin turns to the topic of rudimentary or atrophied organs. Many species have such organs, which are of no apparent use and which often never develop beyond a vestigial stage. Creationists must explain such structures as necessary 'for the sake of symmetry' or 'to complete the scheme of nature', but Darwin once again insists that to invoke a divine plan of nature is no explanation at all. The evolutionist has an obvious explanation since he can show that the rudimentary organs are relics of once useful structures, now declining because the changing habits of the species have rendered them superfluous. Heredity preserves the structure to some extent but it gradually decays because natural selection will favour those individuals who do not waste their energy growing useless structures.

In conclusion Darwin sums up the general implications of his new approach. He looks to the younger generation of naturalists in the hope that they will be able to throw off the prejudices that commit many experienced workers to creationism. He now reveals how far he is prepared to extend his theory. The basic similarities between all living things lead him to infer that 'all the organic beings which have ever lived on this earth have descended from some one primordial form, into which life was first breathed' (p. 484). The latter phrase seems to imply that the original creation of life was by divine miracle. It is not at all clear that Darwin really believed this, but he knew that biologists such as Pasteur had demolished the claim that life was now being produced by spontaneous generation. Perhaps conditions were different in the early periods of the earth's history, but Darwin had no interest in trying to explore the ultimate origins of life. Spontaneous generation was seen as an inherently materialistic hypothesis and he thus felt it safer to leave open the possibility of a supernatural origin.

Darwin expresses the hope that evolutionism will bring about a revolution in natural history as scientists begin to explore the theory's applications. In particular he notes that 'Light will be thrown on the origin of man and his history' (p. 488). This is not the only reference to changes within the human race in the *Origin of Species*, but it is the only absolutely unequivocal statement of Darwin's belief that his theory will account for the origins of mankind from a lower form. He was well aware that the extension of the theory to mankind would provoke controversy because it would threaten the traditional view that our mental powers lift us onto a higher plane than the animals. He hoped to minimize the resulting outcry by refusing to discuss human origins in detail but felt that he had to include at least this brief indication of his beliefs.

To offset this dangerous implication, Darwin concludes with an effort to convince his readers that the theory can be reconciled with traditional beliefs about the relationship between God and nature. He argues that it is better to think of the Creator governing the world by law rather than by arbitrary miracles and suggests that 'as natural selection works solely by and for the good of each being, all corporeal and mental endowments will tend to progress towards perfection' (pp. 488–9). There can be no direct trend leading towards mankind, but we can be sure that the overall effect of natural selection is progressive and hence that the human race is the outcome of a process established by its Creator. Natural selection follows inevitably from the laws of reproduction and is thus the best way for God to ensure progress in so complex a world.

> Thus, from the war of nature, from famine and death, the most exalted object which we are capable of conceiving, namely, the production of the higher animals, directly follows. There is grandeur in this view of life, with its several powers, having been originally breathed into a few forms or into one; and that, whilst this planet has gone cycling on according to the fixed law of gravity, from so simple a beginning endless forms most beautiful and most wonderful have been, and are being, evolved. (p. 490)

8

The Emergence of Darwinism

DARWIN WAS ALREADY a respected scientist but the appearance of the *Origin of Species* turned him into a public figure. From this point onwards it becomes necessary to employ two different perspectives in evaluating his life. On the one hand we have Darwin as a private individual, still chronically ill, living in seclusion with his family and struggling to continue his scientific work. On the other we have Darwin as the public symbol of the controversial new theory of evolution, the figurehead of the movement by which science was seeking to take control of areas of thought once regarded as the province of theologians and moralists. The two perspectives interact, of course, since Darwin took a great interest in the promotion of his theory and kept closely in touch with converts such as Hooker and Huxley who were battling for the idea in the outside world. His own contribution was the writing and revising of his books and the ongoing scientific studies designed to throw light on the nature of the evolutionary process. He also continued to build up the immense communications network that allowed him to draw information from – and to influence – an ever-increasing number of biologists. The enormous number of letters already published, to say nothing of the vast number listed in the *Calendar of Correspondence* awaiting publication, are a tribute to his activity during this period.

At the same time there is a sense in which 'Darwinism' became something other than Darwin's own teaching. In the outside world the concept of evolution was being used both by scientists and non-scientists alike for their own purposes. A few biogeographers, including Hooker and Wallace, continued to develop what might be called a genuinely Darwinian research programme. But many called themselves 'Darwinians' merely because they saw Darwin as the key figure who had initiated the great debate, not because they found his theory of natural selection particularly convincing as an explanation of how evolution worked. Even supporters such as Huxley seem to have had only a limited commitment to those aspects of Darwin's thinking that are seen as most important by modern biologists. Some biologists began to develop openly non-Darwinian theories of evolution based on Lamarckism or the idea of inherently progressive trends. Darwin's great achievement was to force the majority of his contemporaries to reconsider their attitude towards the basic idea of evolution, but he did this despite the fact that many found natural selection unconvincing. We have already seen how the *Origin of Species* was tailored as much as possible to match current expectations. It is hardly surprising, then, that the overall message of evolutionism should to some extent become uncoupled from the detailed logic of Darwin's own position. Any analysis of the impact of Darwin's work must take account of the complex role played by the *Origin of Species* as a catalyst in the transition to late nineteenth-century progressionist evolutionism, an evolutionism that seized upon those aspects of Darwin's writings that modern biologists reject as least original.

The fact that Darwin himself was not altogether converted to what most people at the time would have meant by 'Darwinism' can be seen from the character of his later research. With the exception of his views on human origins (discussed in chapter 10) he chose to develop highly restricted topics that must have surprised many readers unfamiliar with the real logic of his new explanatory system. He had little interest in the growing tendency to interpret the fossil record as the outline of a great progressive sweep up to mankind. Like Voltaire's Candide, he retreated from the great affairs of the outside world and chose to cultivate his own garden. He studied earthworms, the fertilization of orchids, insectivorous and climbing plants. These were not idle or insignificant projects; each helped evolutionism to throw light on what Darwin knew to be a significant problem in natural history. But they stressed an approach to evolution that was very different to that adopted by the naturalists and palaeontologists who saw it as their duty to work out the origin of all the major groups in the progress of life. For Darwin, evolutionism was still to be seen as a way of throwing light on the origin of the adaptive features displayed by particular modern groups. His work made perfectly good sense in the context of his own view of how the theory should be exploited but was widely at variance with what some of his ostensible supporters were doing to promote evolutionism as a world view that would legitimate Victorian pro-gressionism.

Cultivating his Garden

In many respects life in the Darwin household was unaffected by the publication of the *Origin of Species*. Darwin remained chronically ill. He spent six weeks taking the water cure at Ilkley although it seemed to do him little good. When sending Wallace a copy of the *Origin of Species* he complained that he had hardly seen anyone for six months.[1] On returning to Down in November 1859 he was immediately involved with the Magistrates' Court and came home 'utterly knocked up'.[2] He spent most of 1860 at Down and began to work on the *Variation of Animals and Plants under Domestication* which would present his detailed evidence for the existence of a fund of variability on which selection could act. This was to become an extended project; he took an eight-week working holiday at Torquay in 1861 but was again stricken with illness in September 1863 and was incapacitated for six months. Over the next couple of years he gradually improved thanks apparently to a strict diet imposed by Dr Bence-Jones which 'half starved him to death'.[3] This was to be the pattern for the rest of his life; periods of indifferent health in which he could work a few hours a day would be interrupted with occasional relapses in which everything came to a halt.

The family routine established to allow Darwin to cope with his health problems continued as before. There were still children around the house; in 1860 George was fifteen, Francis was twelve and Leonard ten. There were further domestic tragedies. Another son, Charles, had died at the age of eighteen months only a few days after Wallace's paper arrived in 1858. In 1862 both Emma and Leonard became seriously ill with scarlet fever but both recovered. As the surviving children grew up, several began to take a serious interest in science. George was Second Wrangler at Cambridge in 1868 and went on to become Plumian Professor of Astronomy and Experimental Physics.[4] Francis also studied mathematics and natural sciences at Cambridge and in later years helped his father conduct his extensive botanical researches. After his father's death he was appointed lecturer and then reader in botany at Cambridge.

Darwin's favourite relaxation was still having novels read aloud to him by his wife. In the *Autobiography* he complained that there ought to be a law against unhappy endings.[5] He also noted that he had gradually lost his old interest in poetry and paintings and now found Shakespeare insufferably dull. In the end, he concluded, his mind had become a kind of machine for grinding out laws from facts – almost everything except the faculties required for scientific thinking had become atrophied. His love of popular novels suggests that this was something of an exaggeration, but certainly his involvement with science left him no energy for serious reading outside the technical literature.

In the early 1860s his chief concern was the campaign to establish the case for evolution. In announcing the theory of natural selection he had shown that it was possible to break the deadlock created by the apparent lack of any plausible mechanism of change. Obviously he wanted the selection hypothesis to be taken seriously, but it was far more important that the basic idea of evolution become acceptable, whatever the mechanism.[6] In fact natural selection was to remain highly controversial throughout the nineteenth century, but now that the deadlock had been broken the general case for evolution was to become widely accepted within a few years. Darwin had already ensured that he had a few key figures primed to begin the assault on the scientific community. He realized that there would be a flood of opposition from conservative thinkers but hoped that a small band of dedicated followers could withstand the storm long enough for the theory to gain a hearing. Once the possibility of immediate dismissal had been overcome, this small nucleus or fifth column could begin the work of persuasion that would ultimately convert a majority of the scientific community. 'If we can once make a compact set of believers we shall in time conquer.'[7]

Hooker was already a firm convert who was able to write favourable reviews of the *Origin of Species* and provide explicit support for evolution in the Introductory Essay to his *Flora of Tasmania* of 1860. Huxley too had been well primed and received the *Origin of Species* with enthusiasm. Fortunately he was asked to review it for the London *Times*, thus helping to ensure that at least one favourable notice came to the public's attention at a very early stage in the debate. He welcomed natural selection not – as we shall see below – because he was convinced that it held all the answers, but because it showed that new ideas on the question were still possible. His doubts are already apparent in one of his earliest letters to Darwin after the *Origin of Species* appeared: despite all his praise he warned 'you have loaded yourself with an unnecessary difficulty in adopting *Natura non facit saltum* [Nature makes no leaps] so unreservedly'.[8] Throughout the rest of his career Huxley would be tempted by the view that new characters might appear suddenly in response to some power directing the course of variation. Nevertheless he was determined to see that the *Origin of Species* got a hearing. He urged Darwin not to worry about 'the curs which will bark and yelp' because some of his friends were plentifully endowed with combativeness: 'I am sharpening my claws and beak in readiness.'[9]

Towards the end of 1860 Darwin was able to report a number of scientific converts or partial converts. There were four geologists including Lyell, four zoologists, two physiologists and five botanists.[10] Soon, however, the unfavourable reviews began – Darwin was particularly upset by Owen's remarks in the *Edinburgh Review* – and he began to fear a long uphill fight. There seemed to be few new converts and he worried that some supporters would turn back in the face of the attacks.[11] Over the next few years, however, the tide began to turn. In 1864 Darwin was awarded the Copley Medal of the Royal Society, and although there was some debate about the award, we know that Darwin's supporters worked actively to gain him this honour as a symbol of his status as a scientific innovator.[12] By the late 1860s the debate was largely over; so many scientists had converted that there was no longer any possibility of going back. A good illustration of the Darwinians' growing confidence can be seen in a letter from Huxley concerning Hooker's participation in the British Association meeting at Norwich in 1868:

> We had a capital meeting at Norwich, and dear old Hooker came out in great force as he always does in emergencies.
>
> The only fault was the terrible 'Darwinismus' which spread over the section and crept out when you least expected it, even in Fergusson's lecture on 'Buddhist Temples.'
>
> You will have the rare happiness to see your ideas triumphant during your lifetime.
>
> P.S. – I am preparing to go into opposition; I can't stand it.[13]

Evolutionism was now secure, although natural selection was still widely regarded as only a part of the overall mechanism of change.

To a significant extent the promotion of evolutionism depended on the activity of Darwin's early converts (see below). His own contribution consisted of constant encouragement and the continued development and publication of his own ideas. Much of his effort in the period through to the end of 1866 was taken up with the manuscript of the *Variation of Animals and Plants under Domestication* (published in 1868). Here he presented at length, and with appropriate documentation, his evidence for the enormous variability of species. He also included a chapter outlining his theory of heredity, 'pangenesis'. Although supplemented in the 1860s, there seems little doubt that this theory was a distillation of ideas that had been circulating in his mind since the time of the early notebooks.[14] Far from being an anticipation of modern genetics, pangenesis was essentially a theory of generation in the old tradition. Indeed one of the most frequent criticisms levelled against it was that it did little more than revive concepts that had been circulating since the time of Hippocrates. Darwin believed that every part of the body buds off minute particles which he called 'gemmules', each of which has the capacity to grow into the appropriate organ. The gemmules circulate through the body and are collected in the sexual organs. Reproduction takes place when gemmules from both parents are mixed together in the fertilized ovum, and the embryo grows through the developmental power of the gemmules themselves.

Apart from its apparent lack of originality pangenesis gained few converts and was soon overtaken by developments in cell theory which rendered it most implausible that gemmules could exist independently of cells. The theory's role in the development of evolution theory has provoked much debate among historians. In the accounts of his father's work provided in the *Life and Letters* and *More Letters*, Francis Darwin minimized the significance of pangenesis because he was aware that it represented one of the least enduring parts of the overall theoretical system. This has led one modern historian to complain that Francis provided a distorted image of his father's interests, concealing the fact that he had always seen evolution as an interaction between the reproductive process and the external environment.[15] Pangenesis shows that the early interest in the role of sexual reproduction was sustained throughout Darwin's life, influencing his thinking on the way natural selection could work even in the post-*Origin of Species* period.

But exactly what did this rather conventional theory of reproduction imply for the workings of the evolutionary mechanism? There is a widespread assumption that Darwin's 'failure' to adopt a particulate view of heredity in anticipation of Mendelian genetics constituted the major barrier to his theory's plausibility. Pangenesis supposed that each character of the offspring was built from a number of gemmules derived from the appropriate part of both parents' bodies. Thus it implied that heredity was a process in which parental characters were blended together, not transmitted as unchangeable units as Mendel was to demonstrate. According to the interpretation adopted by Loren Eiseley, blending heredity would render natural selection unworkable.[16] As the engineer Fleeming Jenkin pointed out in an 1867 review of the *Origin of Species* a favourable new character would be swamped by blending because its effect would be halved in each successive generation due to interbreeding with unchanged individuals.[17] Like a spot of black paint put into a bucket of white and stirred, its effect would soon disappear altogether. Eiseley suggests that selection could only be seen as an effective mechanism if it were supposed that the favourable new character was inherited as a unit and was thus immune to dilution in this way.

He argues that in the face of Jenkin's critique Darwin himself began to abandon the selection theory and turn instead to Lamarckism. Pangenesis allowed for the inheritance of acquired characteristics because the parents' bodies actually manufacture their gemmules rather than merely transmitting unit characters as in the modern concept of the gene.

More recent historians have tended to reject Eiseley's view both of the impact of Jenkin's review and of the overall implications of pangenesis. In fact natural selection *would* be able to work if heredity was a blending process, provided favourable variations are plentiful. Jenkin had been thinking of single individuals with significantly new characters – monstrosities or 'sports of nature'. Darwin took the review seriously not because he himself thought that evolution worked in discontinuous steps but because he did have doubts about how plentiful the supply of even small variants would be. Some time before Jenkin's review appeared, Wallace was urging him to insist that for most characters there is a range of variation already existing within the population.[18] In the case of height in the human population, for instance, there is a wide range from the tallest to the shortest, with the majority of people clustered fairly closely around the mean value. On such a model of variation, blending is no problem because if, for instance, tall people were at an advantage in a certain environment there would always be a large number of them for selection to act upon. The advantage would be gained by everyone of above average height, not just by a few exceptionally tall individuals. The claim that natural selection was rendered implausible by blending is a myth created by modern biologists whose viewpoint has been shaped by hindsight. Because we now know that Darwin's theory of heredity was wrong in that it did not anticipate Mendelian genetics, it is all too easy to fall into the trap of assuming that this must be the key factor that limited acceptance of the theory in his own time. In fact heredity was but one of the many objections raised against natural selection, and we shall see that there were far more powerful forces at work generating an interest in non-Darwinian evolutionary mechanisms.

At the same time that he was working on his material on variation and heredity, Darwin was also instituting new research projects designed to throw light on evolution. Although he never called himself a botanist, many of these later projects involved studies of plant characteristics that were intended to show how evolution could have produced them. Down House had extensive gardens and in 1863 a new hothouse was built. Francis Darwin, who became his father's assistant in this long series of experiments, recalled how he would persevere despite the considerable difficulties – counting plant seeds by simple microscope, for instance, in extensive breeding trials. Darwin's motto was 'It's dogged as does it' and over the years his efforts were rewarded with a series of important publications.[19]

The first of these botanical projects was a study of the mechanisms by which flowers are fertilized, especially in the orchids. Darwin worked actively on this during 1860 and 1861, preparing a paper for the Linnean Society and his book *On the Various Contrivances by which British and Foreign Orchids are Fertilized by Insects* (1862). The term 'contrivance' is significant here: Darwin was attempting to demonstrate the complexity of the mechanisms that have evolved so that insects will transmit pollen from one plant to another, thus ensuring cross-fertilization. Francis Darwin wrote that his father had revived teleology in the study of natural history, by which he meant that the work on orchids showed that even the most complex flowers had an adaptive purpose.[20] Darwin also showed that these complex mechanisms could have been produced by natural selection because there are usually other species with less well-developed forms of the complex character illustrating the route that evolution might have taken. The insects too have evolved to match the floral structures since it is important for them that they are able to reach the nectar provided by the flowers. In effect Darwin was asking his readers if they could believe that such a complex set of structures and interactions could have been produced by a Creator individually designing each species and its associated insects. Clearly he himself found this possibility implausible and saw his demonstration of the complexities of adaptation as indirect support for evolution by a mechanism such as natural selection.

The explanation for the orchids' structures rested on the assumption that it was to the plants' advantage to be fertilized by pollen from another individual, avoiding self-pollination if possible. Darwin had of course been interested in the process of fertilization since the 1830s, but his study of cross-fertilization began as the result of a chance observation that, of two beds of *Linaria vulgaris*, the crossed plants seemed much more vigorous.[21] Darwin became convinced that this extra vigour was so important that plants would develop mechanisms to ensure that crossing would take place. From that chance observation flowed eleven years of careful experiments intended to demonstrate the truth of this assumption, culminating in his book *The Effects of Cross and Self-Fertilization in the Vegetable Kingdom* (1876). He also studied those species in which there is more than one type of flower, such as the primrose, and showed that this phenomenon was also linked to the need for cross-fertilization. Another book resulted from this study, *The Different Forms of Flowers on Plants of the Same Species* (1877).

Darwin also conducted a series of experiments on the movements of plants, especially the climbing ability possessed by some species. Here again he was concerned to show that the power to climb was an adaptation that could have been developed by natural selection. He investigated the mechanism by which hops and other plants climb by twining around supporting poles and the ability of other species to grasp onto supports by means of tendrils. The fact that so many different groups of plants have acquired the ability to climb suggested to Darwin that all plants must possess some rudimentary power of movement. A paper on this topic was read to the Linnean Society in 1865 and subsequently modified to form the book, *The Movement and Habits of Climbing Plants*, published in 1875. He also published another book, *The Power of Movement in Plants* (1880), showing that the twining of shoots was linked to the fact that the tip was sensitive to light and some-how influenced growth lower down the shoot to produce a bending effect.

In 1860 Darwin had become interested in the ability of the common sun-dew, *Drosera*, to catch and digest small insects. He undertook a long series of experiments to test how this and other insectivorous plants caught and digested their prey. An illustration of his interest – and of the plants' sensitivity – can be seen in a letter to Hooker: 'I have been working like a madman at Drosera. Here is a fact for you which is as certain as you stand where you are, though you won't believe it, that a bit of hair 1/78000 of one grain in weight placed on a gland, will cause *one* of the gland-bearing hairs of Drosera to curve inwards, and will alter the condition of the contents of every cell in the foot-stalk of the gland.'[22] Darwin was convinced that the ability was an important adaptation designed to supply the plants with nitrogen and thus to allow them to grow on extremely poor soil. His book on *Insectivorous Plants* was published in 1875.

Finally there came an extended study of the activity of earthworms. Darwin had been interested in the process by which worms affect the soil ever since his days as a geologist in the 1830s. In 1871 a trench was driven across a field near Down which had been covered with broken chalk in 1842. The chalk was now found to be buried to a depth of seven inches by soil brought up to the surface by worms during the intervening years. Darwin began to study the quantity of worm castings brought to the surface in each year and calculated that over an acre the total amount raised was eighteen tons. He was convinced that the worms' activity was vital to the production of the vegetable mould which supports plant life. They eat decayed leaves and also process the soil through their bodies, thus both fertilizing and aerating the soil. The lowly earthworm was thus a vital factor in soil ecology and a significant geological agent, altering the very face of the landscape over a long period of time.

The Formation of Vegetable Mould through the Action of Worms was published by John Murray in 1881, the year before Darwin's death. It was quite a success: 'My book has been received with almost laughable enthusiasm, and 3500 copies have been sold.'[23] He complained of being 'plagued with an endless stream of letters on the subject'. Clearly Darwin had hit upon a topic that attracted the attention of the garden-loving public. Yet a study of earthworms was hardly what most Victorians would have expected from the founder of evolutionism. Much of Darwin's later scientific work had taken him away from the great issues of the day to concentrate on small-scale topics that could be illuminated by his particular approach to evolutionism. Far from writing a great survey of the history of life on earth he chose to investigate the origins of particular adaptations in the light of his theory of natural selection. He was not uninterested in the larger dimension of evolutionism – a letter to Lyell in 1860 offered two possible lines of ancestry for the mammals and made it clear that he preferred a divergent rather than a step-by-step model of development.[24] But Darwin knew that in most cases the fossil evidence to decide such questions would be lacking and he chose not to comment in public on the major steps in the ascent of life.

Where many 'Darwinians' saw the reconstruction of the history of life on earth as their major goal, Darwin himself held back because he knew that his particular theory of evolution had more to offer when applied to a different kind of question. He had never approached the topic from the study of the fossil record and in this respect his whole technique was quite different to that of many nineteenth-century evolutionists. Darwin preferred to study the modern world in an effort to elucidate the processes that have governed the formation of the most recently evolved species. Extending evolutionism back into the remote past created a different kind of problem. Such problems could be – and still are – studied by palaeontologists seeking to use fossil evidence to piece together a plausible sequence to explain the origin of major living groups. But Darwin's theory was created from the study of small-scale changes in the modern world. It was a theory that concentrated on explaining the actual processes that still affect every living thing, not on the reconstruction of long-past evolutionary links. In this respect Darwin's later botanical studies were characteristic of his own personal approach to the problems of evolutionism – but they ignored what many of his contemporaries would have seen as the most important questions in the study of the progress of life on earth.

The Darwinians

In the course of the 1860s the efforts of Huxley and other early supporters were enough to bring about a revolution leading to the general acceptance of evolutionism. From an attitude of hostility, or at best of 'wait and see', the majority of biologists were converted to open support for the basic idea that new species originated from old ones by a process of transmutation. The *Origin of Species* had clearly played a major role in precipitating this change, and many evolutionists chose to call themselves 'Darwinians' or 'Darwinists' in acknowledgement of the fact that Darwin had led them to confront this new area of biology.

Within a decade or two, opponents such as Samuel Butler were complaining that the Darwinists had taken control of the scientific community and established a new dogmatic orthodoxy that suppressed any attempt to question its basic assumptions.

But who were these Darwinians? What did they believe and how did they engineer this transformation? Modern historians now suspect that the answers to such questions are by no means simple. Some Darwinians followed research programmes stemming from the major innovations that Darwin had suggested. But others took up just those aspects of evolutionism that Darwin himself had ignored, and in some cases they seem to have had doubts about the efficacy of natural selection as a mechanism of evolution. There were major objections to the selection theory that were never overcome in Darwin's lifetime, and to some extent his followers had to argue around these problems rather than solve them. Open criticism of the selection theory grew in intensity during the last decades of the century and explicitly anti-Darwinian versions of evolutionism were established. If we are to understand the scientific revolution that Darwin initiated, we must move beyond the simple assumption that his theory triumphed because an overwhelming body of scientific evidence was immediately brought forward to substantiate it.

One modern historian has gone so far as to suggest that it is impossible to define a rigid conceptual foundation for nineteenth-century Darwinism.[25] A Darwinist was someone who expressed personal loyalty to Darwin – and in many cases the views of what we might call 'pseudo-Darwinians' were not so very different from those of his opponents. This suggestion is probably a little too drastic but it does help to focus attention on the fact that the advent of Darwinism was a social event within the scientific community and must be understood in terms of changing loyalties as well as changing research programmes. Most of Darwin's opponents were concerned about those aspects of his theory which tended to undermine the old belief that nature was a divinely planned structure. They were willing to accept evolutionism but only if they could believe that it represented a process with a structure and a goal that was imposed on it by God. Some pseudo-Darwinians were also unwilling to accept Darwin's totally open-ended view of the evolutionary process but they *were* willing to see the direction of change as being under the control of purely material forces. In the end the success of Darwinism rested not on a general acceptance of the selection theory but on the exploitation of evolutionism by those who were determined to establish science as a new source of authority in Western civilization.

Although the Darwinians shared a commitment to scientific naturalism, they did not all come from the same scientific background and at a technical level they did not accept the theory for the same reasons. Those who came closest to Darwin's own position were biogeographers such as Hooker and Wallace. The geographical distribution of species provided the clearest evidence for adaptive evolution and thus indirectly supported Darwin's claim that natural selection was the chief mechanism of change. But many biologists were not conditioned to think along these lines. Huxley, for instance, was essentially a morphologist rather than a field naturalist. He was trained to study the structure of living and fossil species in an effort to discover the underlying similarities between them. Morphologists had little interest in how animals and plants adapt to changes in their local environment or in mechanisms of geographical dispersal. From their perspective, evolutionism offered the prospect of translating the abstract 'relationships' used to classify animals and plants into real links via hypothetical ancestries. Fossils and living species would be arranged into the most plausible evolutionary 'tree', but the shape of this tree would be reconstructed from abstract comparisons that paid little attention to the practical realities of adaptation or to the geographical dimension of evolution. The morphological tradition was firmly established in pre-Darwinian biology and would survive the transition to evolutionism, but it remained a fertile soil within which both pseudo-Darwinian and non-Darwinian ideas would flourish.

From the morphologists' perspective there were a number of problems with natural selection and it was only a matter of emphasis which determined whether an individual biologist would become a pseudo-Darwinist or an outright opponent of Darwinism. It was easy for a student of animal form in the abstract to imagine that some characters were not shaped by adaptive pressures at all, and equally easy to imagine that the transition from one species to another might be instantaneous, that is, by saltation rather than gradual transformation. Many morphologists – Huxley included – were attracted to the idea that variation might not be random (as Darwin supposed) but might be led along a predetermined course to give a neat linear pattern of evolutionary development. It was all too easy to imagine that natural selection merely weeded out those lines of evolution that were hopelessly maladaptive, while the true source of new developments was some predetermined trend that would give evolution an artificially regular pattern. From such a position it was only a short step to deciding that natural selection was really only a secondary or negative mechanism of evolution, with positive changes being the result of some non-Darwinian force directing variation.

The true Darwinists are easy to spot because they retained an interest in those problems that had encouraged Darwin himself to develop his unique perspective. Hooker and Wallace, for instance, had both been converted to evolutionism by the evidence of biogeography. After several years of debate with Darwin, Hooker had been brought around to see the advantages of evolutionism in time to lend immediate support to the *Origin of Species*. He continued to study the geographical distribution of plants with a view to understanding the role of migration in the evolutionary process. Darwin and Hooker continued their debate on the mechanisms of dispersal through into the post-*Origin of Species* period, Hooker preferring to invoke land 'bridges' between continents which had now sunk beneath the ocean while Darwin preferred accidental mechanisms such as transportation by driftwood rafts and storm winds.[26]

Wallace too became a staunch 'Darwinist' and – whatever the limitations of his initial conception – a strong exponent of natural selection. He published extensively on the relationship between species and varieties and on the problems of biogeography. He also corresponded frequently with Darwin on a host of issues connected with the operations of natural selection, including the build-up of interspecific sterility and sexual selection.[27] As field naturalists and explorers, Hooker and Wallace were in a position to appreciate how evolution might depend on the hazards of migration and on how small isolated populations adapted to their local environments. They could certainly show how evolution might illuminate problems on a global scale as geological changes gradually shaped the world within which modern species must live. The boundary between the Australian and Asian faunas in what is now Indonesia is still called 'Wallace's line' because Wallace was the first to identify it and to explain it in terms of the possibilities of migration in earlier periods with a lower sea-level. But field naturalists had no incentive to study remote problems such as the origin of new classes and no reason to suppose that evolution might be governed by long-range trends that would drive species in predetermined directions whatever the environments to which their members might be exposed.

Huxley is a far more difficult figure to characterize. At first sight he appears to be a committed Darwinist – he even became known as 'Darwin's bulldog' because of his vigorous counter-attacks against the theory's opponents. He later recalled that his first reaction on reading the *Origin of Species* was 'How extremely stupid not to have thought of that!'[28] Yet recent research has suggested that Huxley's commitment to the selection theory was at best only lukewarm.[29] He was a staunch advocate of scientific naturalism who had refused to commit himself to evolutionism only because he saw no plausible hypothesis on the mechanism of change. Natural selection was important as a working hypothesis that demonstrated the scientists' ability to penetrate this hitherto uncharted area, and on these grounds Huxley was determined to ensure that Darwin got a fair hearing. But he had substantial reservations about the theory that Darwin was never able to overcome. We have already noted that from the start he argued that evolution might sometimes work by dramatic saltations rather than by the selection of everyday variations. He also suggested that variation might be directed along fixed lines, with selection merely eliminating those trends that started to go off in a harmful direction.[30] There is nothing to suggest that Huxley ever became a Darwinist in the sense that he took on a research programme that was inspired by the detailed theory that Darwin had proposed.

Huxley is, in fact, a classic example of a pseudo-Darwinian. He accepted evolution because of his enthusiasm for naturalistic explanations, not because he appreciated the real logic of the Darwinian theory. Huxley was a formally trained morphologist, not a field naturalist in the tradition from which Darwin sprang. The difference in their outlook can be seen despite the fact that Huxley, like Darwin and Hooker, gained his reputation in part as the result of a voyage of

PLATE 13 *Thomas Henry Huxley* in 1857; photograph used as the frontispiece to *The Life and Letters of Thomas Henry Huxley*, vol. II

discovery. He spent the years 1846–50 aboard HMS *Rattlesnake*, mostly in Australian waters. But, unlike Darwin, Huxley did not develop an interest in geographical distribution; his reputation was based on his description and classification of newly discovered marine creatures. If he could use the idea of evolution it would be to give a naturalistic basis to the relationships he was trained to see between the various kinds of living structures.

Even the basic idea of evolution seems to have played no role in Huxley's palaeontological work during the early 1860s. For all his vigorous defence of Darwin he made no effort to use the search for evolutionary ancestries to throw light on the fossils he was describing. Only in the late 1860s did he begin to search actively for 'missing links' in the fossil record that would help to provide evidence for evolution. The stimulus which seems to have precipitated this change of attitude was his reading not of Darwin but of the German evolutionist Ernst Haeckel's *Generelle Morphologie* of 1866. Although Haeckel was inspired to become an evolutionist by his reading of the *Origin of Species*, he had little understanding of natural selection and offered a theory that can be seen as a classic expression of nineteenth-century developmentalism. It was Haeckel who linked 'Darwinism' to the claim that the growth of the modern embryo recapitulates the evolutionary history of its species. And, just as the embryo grows in an apparently purposeful manner towards maturity, Haeckel pictured evolution as a tree with a central trunk running directly through to mankind at the head of creation. Progress towards the human form was nature's central goal; all other developments were mere side branches. When popularized in English translation, Haeckel's *The History of Creation* (1876) and *The Evolution of Man* (1879) helped to create the impression that evolution was a process of necessary development towards a single goal, and that the central purpose of the evolutionary biologist must be to reconstruct the main steps in the ascent from the study of living and fossil species.

The fact that Huxley should have been inspired by Haeckel's largely non-Darwinian version of evolutionism must force us to reassess his commitment to Darwinism. In the late 1860s Huxley certainly became actively involved in the attempt to verify evolution using newly discovered fossils. He was one of the first to appreciate the significance of *Archaeopteryx*, the reptile–bird intermediate from the Jurassic rocks of Bavaria that was one of the earliest 'missing links' to be discovered. He also took the lead in promoting the view that fossils could illustrate the complete pattern of development by which modern species had evolved, describing new discoveries of fossil horses from America as 'demonstrative evidence of evolution'.[31] But such attempts to use fossils as illustrations of the path taken by evolution were not representative of Darwin's own approach to the question. There is no doubt that Darwin saw the relevance of Huxley's efforts – the fossil record was too important for evolutionists to ignore – yet the attempt to reconstruct the whole pattern of life's development on earth was not Darwin's chief concern. As a scientist Huxley was a pseudo-Darwinist: his real concern was evolutionism, not the selection theory, and he was inclined to prefer a model for the evolutionary process that owed more to non-Darwinian influences.

If this is so, how can we explain the dramatic transformation that Darwin's book was able to bring about within the scientific community? If his supporters consisted of a medley of biologists some of whom could not even appreciate the main implications of the selection theory, why were they so successful in precipitating a general transition to evolutionism? The conventional image of the debate is centred on highly visible confrontations such as the one at the British Association meeting in Oxford in 1860, at which Huxley is popularly supposed to have demolished the anti-evolutionary arguments of Bishop Samuel Wilberforce. Scientific rationality is supposed to have demonstrated its superiority over traditional superstition. We now know that this image is a false one created by the supporters of scientific rationalism to bolster their own interpretation of the past in which science is ever triumphant in the 'war' against religion.[32] In fact, Huxley did *not* convince the majority of people in his Oxford audience, and the general conversion to evolutionism was not completed for some years.[33] To explain what was going on, historians are now looking beyond the evidence for evolution to the social pressures that were at work within the scientific community and within Victorian culture as a whole.

The new interpretation suggests that Darwin was able to initiate a scientific and cultural revolution because he linked his own very specific interests in evolutionism to a more general trend in Victorian intel-lectual life, a trend that reflected the changing power structure of British and indeed of Western society. Against a background of ongoing social unrest, the middle classes whose wealth was derived from the new industrialization were seeking to wrest control of society from the old landed interests. Science was an important battleground because any challenge to the authority of scripture threatened to undermine the conceptual foundations of the establishment's claim that the existing structure of society was divinely preordained. Evolutionism was an important scientific innovation because it could be used to suggest that nature was an inherently progressive system. Social progress could be seen as a continuation of natural evolution, the inevitable replacement of outdated forms by those more advanced. The inevitability of progress should reassure everyone that what was going on would ultimately be for the benefit of all.

Darwin and the majority of his followers came from a class which saw evolutionism as a means of demonstrating the superiority of new ways of looking at nature and society. But for men like Hooker and Huxley there was a more personal dimension to the struggle. They reflected the new sense of professional identity emerging as scientists and engineers demanded that their expertise be recognized as an important contribution to industrial development. Huxley was acutely conscious of this factor because he had great difficulty obtaining a position after his return from the voyage on the *Rattlesnake*. He was taken on as a palaeontologist at the Royal School of Mines just in time to establish himself before the appearance of the *Origin of Species*. Evolutionism was important to him because it expanded the realm of scientific explanation and thus supported the claims of professional scientists to be taken more seriously. The scientists who studied natural law were to become the new source of intellectual authority, taking over from the moralists and theologians who had once dictated how human nature was to be understood.

We have already explored Darwin's origins within the radical tradition, and we should see his efforts to build up a network of supporters both before and after the publication of the *Origin of Species* as a deliberate attempt to create the foundations for a reorganization of the scientific community that would force everyone to come to grips with the issue of evolution. Earlier in the century the ideas of Lamarck and Chambers had been discredited by biologists loyal to the old tradition, and evolutionism had been branded as materialistic and revolutionary. The new initiative would require a deliberate effort to reconstruct this image; evolutionism might be materialistic but its emphasis on the purposefulness of nature meant that it did not threaten to sweep away the whole foundation of traditional thought. Middle class scientists such as Darwin and Huxley needed to restate the case for evolution in a way that would allow them to maintain the attack on creationism while reconstituting the theory as a basis not for revolution but for gradual progress under their own control.

Darwin had clearly recognized that the unique perspective he had gained from his studies in biogeography and animal breeding offered the new initiative that his fellow scientists were looking for around which to build their case. He carefully built up his contacts with those biologists whom he saw would be most likely to welcome a new initiative, including even those such as Huxley who were not in a position to appreciate the detailed arguments for natural selection. The glue that would hold the supporters together, despite their different scientific interests, was the belief that natural developments were governed by law rather than divine predestination. By presenting evolution as a process governed solely by the normal laws of nature they could imply that social progress was the result of individual human efforts, the centrepiece of the liberal philosophy. Darwin's theory was not a simple projection of that philosophy onto nature since it had been filtered through a unique collection of scientific studies that raised grave problems for the idea of progress. But its emphasis on law and purpose did help others to translate their faith in progress into a general theory of evolution, even though the resulting theory was often rather different from that which Darwin himself had put forward.

It is easy to see why a biogeographer such as Hooker appreciated both the general case for evolution and the detailed arguments for natural selection. In the case of Huxley we have a younger scientist who was clearly ambitious to make his way in the new kind of society but whose training was in a field that made it difficult for him to appreciate the logic of selectionism. However, he was aware of the extent to which the study of animal form had become sterile without the basic idea of evolution and was on the lookout for any new idea that would help him to make the case for a reorganization. Thanks to Darwin's influence, he welcomed selectionism not as the foundation for a restructuring of his own research but as a new idea that could break the deadlock into which the case for evolutionism had been forced by the discrediting of Lamarckism. He was thus willing to fight on Darwin's side even though he had major reservations about the selection mechanism itself. When he finally began to use the idea of evolution in his palaeontological work, his real inspiration was Haeckel's largely non-Darwinian campaign to reconstruct the history of life on earth. Huxley's decision to promote a thoroughly progressionist version of evolutionism may also have been prompted by his recognition of the growing social tensions of the 1860s. In his campaign to persuade working men that their interests lay more with reform than with revolution, the inevitability of evolutionary progress offered an excellent model on which to base his image of social development.[34]

What passed for 'Darwinism' in the 1860s was a combination of two different programmes, one based on a direct expansion of Darwin's own approach, the other on the exploitation of a developmental model of evolution that was, in the end, more in tune with the progressionist world view of the Victorian middle classes. The important point is that the advocates of the two approaches maintained their loyalty to the Darwinian figurehead and remained united by their common faith in naturalism and in the liberal philosophy of progress. There were still many scientists and non-scientists with more conservative inclinations and the arguments for evolutionism were not overwhelmingly convincing. In these circumstances it was important for the supporters of the new idea to play the game of scientific politics very carefully.

Darwin had made a good start by building up a network of contacts who had been prepared to receive the new idea and he now depended on them to fight his battles both in the public arena and in the 'behind the scenes' activities of the scientific community in which new policies were decided. Fortunately he had chosen followers who were particularly adept at playing the political game. Huxley and Hooker were members of the informal 'X club' which exerted considerable influence on the scientific community.[35] Their different interests in evolutionism might have led them to dispute with one another in public, but instead they maintained a united front against the common enemy and worked tirelessly to ensure that evolutionary papers would be published and that scientists favourable to their cause would have access to research funding and academic appointments. It was by playing this game – not by fighting bishops in public – that Huxley fulfilled the expectations that Darwin must have had when he recruited him. Modern scientists may be reluctant to admit that the success of a new theory rests on the public-relations skills of its early supporters, but there can be little doubt that Darwin's initiative succeeded (where it could very easily have failed) because he had already planted the seeds of a political revolution within the scientific community. Having seen the revolution succeed, he then devoted himself to a series of detailed studies that made sense within the logic of his own approach but which were far removed from the global progressionism that Haeckel and Huxley were promoting.

Not that Darwin ignored his friends in their ongoing battles within the scientific community. In 1872 Hooker's position at Kew Gardens was threatened when he was subordinated to an unsympathetic Commissioner of Works appointed by Gladstone's government. Darwin and Huxley were active in the campaign to regain Hooker the autonomy he needed to conduct his scientific work unmolested by officialdom.[36] A year later Huxley became ill with overwork and the following letter from Darwin illustrates the way in which his friends rallied round.

My Dear Huxley – I have been asked by some of your friends (eighteen in number) to inform you that they have placed through Robarts, Lubbock & Company, the sum of £2100 to your account at your bankers. We have done this to enable you to get such complete rest as you may require for the reestablishment of your health; and in doing this we are convinced that we act for the public interest, as well as in accordance with our most earnest desires. Let me assure you that we are all your warm personal friends, and that there is not a stranger or a mere acquaintance amongst us. If you could have heard what was said, or could have read what was, as I believe, our inmost thoughts, you would know that we all feel towards you, as we should to an honoured and much loved brother. I am sure that you will return this feeling, and will therefore be glad to give us the opportunity of aiding you in some degree, as this will be a happiness to us to the last days of our lives. Let me add that our plan occurred to several of your friends at nearly the same time and quite independently of one another.
– My dear Huxley, your affectionate friend,

Charles Darwin[37]

Huxley recovered and went on to become one of the most respected public figures of the late nineteenth century. Later on, Darwin was also instrumental in obtaining a public pension for Wallace in his declining years.[38]

One can perhaps see why the opponents of Darwinism complained that the movement had gained a stranglehold on the scientific community. The Darwinians formed a tightly-knit group held together by personal loyalties and commitment to a particular ideology. It was *not* held together by a shared research programme, since the true 'Darwinism' based on biogeography and the study of adaptive evolution had few points of contact with the morphological approach of pseudo-Darwinians such as Huxley. Yet the commitment to a belief that nature was governed universally by the operations of natural law held the group together, allowing them to present a united front even when their scientific work did not mesh very well together. Darwin's great triumph was that he had used his own unique approach to evolution as a catalyst that had enabled the exponents of progressionism to transform Victorian thought. Although his own vision of evolution as a haphazard process driven by the pressures of local adaptation had little to offer those who sought to reconstruct the ascent of life on earth, the appearance of a new mechanism of change had turned the balance in the general debate over the plausibility of natural development.

Those who opposed Darwinism were not diehards who wished to retain a purely biblical view of creation. Many of them were willing to accept the general idea of evolution and adapt it to their own beliefs. But on the whole they were suspicious of the ideological agenda that lay implicit in the Darwinians' appeal to the universal efficacy of natural law. They objected to the image of haphazard development at the heart of Darwin's theory because they wished to retain the view that nature was in some senses the expression of a divine purpose and because they did not believe that progress was merely the summing up of a vast multitude of trivial everyday occurrences. There were many scientific arguments against evolution but underlying most of them was a desire to resist the Darwinians' assumption that evolution could be used as a model for the liberal view of progress favoured by the middle classes. Some of the arguments could be well appreciated by a pseudo-Darwinian such as Huxley, since they often reflected the underlying values of the morphological tradition in biology. But the Darwinians can be distinguished from their opponents quite clearly on the question of design or purpose in the universe. Even a pseudo-Darwinian such as Huxley wanted to use evolutionism as a means of rejecting the traditional view that nature can only be explained as an expression of a higher Power whose intentions are fulfilled by the pattern of evolutionary development. The opponents *did* wish to retain this view and they were prepared to marshal an impressive battery of arguments to defend their alternative image of evolutionism.

9

The Opponents of Darwinism

<hr>

DARWIN REALIZED THAT by reopening the case for transmutation he would at last have to face the opposition he had avoided for so long. Things had changed since the 1840s but there were still many scientists and religious thinkers prepared to resist this last challenge to the old theological view of nature. Conservative naturalists brought a battery of technical arguments to bear on the theory, forcing Darwin to respond in a variety of ways including modifications to successive editions of the *Origin of Species*. Nor was there any shortage of opponents willing to point out the theological and moral shortcomings of the new theory. Darwin was able to survive this onslaught primarily because he had prepared the way by building up a small but influential network of naturalists willing to use his idea as an excuse for a renewed assault on the question. But it was touch and go for the first few years; under slightly different circumstances it would have been possible for the selection theory to have fizzled out. Far from being a self-evident solution to the problem of evolution, natural selection was a highly original explanation that was never free from criticism throughout Darwin's lifetime. The Darwinists could not simply sweep the opposition aside by sheer force of argument. They had to fight a political battle both within and outside the scientific community, in the course of which a public image of Darwinism emerged that was by no means an accurate projection of what Darwin had proposed.

Historians of the Darwinian Revolution have often tended to over-simplify the debate by concentrating solely on the Darwinists' successful campaign of the 1860s. Modern Darwinists who take an interest in the origins of the theory almost inevitably tend to assume that the early opposition was essentially short-sighted. They are convinced that Darwin had, in fact, come up with a successful solution to the question of how species evolve, and have little patience with those of his contemporaries who could not appreciate this. If modern biologists are willing to concede any value to the anti-Darwinian arguments, it is in the area of heredity where we know that Darwin's theory of pangenesis failed to anticipate the particulate model subsequently introduced by the Mendelians.[1] Modern opponents of Darwinism – and there are many of these, at least outside the scientific community – take a different view of the situation. For them, Darwin's theory was fundamentally flawed from the start, and they see the opponents as the source of valid objections that can still be taken seriously today. Darwin's response to the criticisms is presented as an inadequate attempt to shore up a fundamentally unsound structure.[2] All too often these modern opponents imply that the early success of Darwinism was due merely to its ability to uphold the materialistic ideology of Victorian capitalism.

We have already seen that ideology *was* a factor in the debate: Darwin's theory was exploited by liberals such as Huxley who wanted to challenge the conservative domination of the scientific community. But to dismiss the selection theory as nothing but an expression of Victorian capitalism is to miss the fact that pseudo-Darwinism evaded what many modern biologists see as the most important implications of Darwin's theory. Such an interpretation also depends upon a convenient refusal to acknowledge that the opposition to Darwinism, far from diminishing as the century progressed, actually grew in strength after the 1870s. In these circumstances we must be very careful indeed when we evaluate the scientific arguments that were advanced for and against the selection theory.

To achieve a more balanced perspective we need to appreciate the underlying issues at stake in the controversy. There were indeed a number of scientific difficulties that could be raised against the selection theory. These were not confined to the area of heredity – indeed, the temptation to concentrate on Darwin's 'failure' to anticipate genetics gives us a picture of the debate that is distorted by hindsight. His opponents were far more interested in arguments that seemed to undermine the case for branching evolution brought about solely in response to local environmental pressures. What they really objected to was the unpredictability of a system in which the history of life could be influenced by the hazards of migration and the multitudinous possibilities for how a population might adapt to a new environment. The opponents preferred to believe that history was predetermined so that it could only unfold in accordance with a predictable pattern. Their arguments need to be taken seriously because they reflect underlying attitudes that are characteristic of nineteenth-century biology (however outdated they may seem today). Even Huxley was a morphologist who saw evolution mainly in terms of abstract lines of descent, an approach that bore little relationship to the kind of detailed studies undertaken by Darwin. If Huxley had problems with natural selection, how much more powerful must those objections have seemed to naturalists who did not share his commitment to the idea of development under natural law?

To understand the motivations of the opponents we must accept that their misgivings reflect a scientific tradition that differs significantly from the programme that Darwin was trying to establish. The biogeographical perspective was only a small part of late nineteenth-century natural history and tended to be swept aside by the desire to find all-embracing patterns encompassing the whole fossil record. And in the period before the emergence of Mendelian genetics there were many other hypothetical mechanisms of evolution that seemed more plausible than natural selection. It was not so much that the lack of genetics rendered selection unworkable, but that the pre-genetical view of variation and heredity – which Darwin himself accepted – allowed all sorts of other mechanisms, including Lamarckism, to flourish. Instead of concentrating exclusively on the objections to natural selection we need also to take account of the alternative concepts of evolution that the opponents preferred. This will give us a far more accurate picture of late Victorian evolutionism and will allow us to understand what it was about Darwin's theory that failed to mesh with the expectations of the time.

The overall success of Darwin's campaign can be measured by the fact that the opponents were increasingly compelled to present alternative theories of how evolution worked. In effect they had given up creationism and conceded the basic point that new species were the transformed products of old ones. Their objections focused on Darwin's particular explanation of how the process worked. The nature of their objections and of their preferred alternatives show us that the real problem with natural selection was its challenge to the belief that the development of life had a structure revealing an underlying divine purpose. To rely on the selection of random variants by environmental pressure was to descend into pure materialism. Only by seeing regular patterns or some other sign of purpose in nature would it be possible to preserve traditional beliefs. The scientific arguments reflect underlying religious concerns, but these in turn reflect the ideological debate that was transforming the Victorian world. Huxley had his scientific doubts about the adequacy of natural selection but he was prepared to go along with Darwinism because he was committed to the view that progress must be the product of the everyday actions of natural forces. For many opponents it was precisely that underlying assumption that was open to question. Evolution might occur, but if it was to be seen as the unfolding of a meaningful divine plan, there must be something visible in the process that could not be reduced to everyday events.

In these circumstances the debate was bound to be a complex affair. Many older scientists were reluctant to break completely with the tradition of natural theology, even though they accepted the idea of evolution. Some theologians welcomed evolutionism – as long as they could see it in their own light. Some naturalists were committed to the ideological programme of Darwinism even though they found natural selection hard to swallow. Social evolutionists were able to promote their ideas of progress using models that were often only superficially Darwinian and which owed at least as much to the older tradition of Lamarckism. Only by recognizing the complex relationship between Darwin's actual scientific theory and the various religious and ideological positions of the time will we be able to understand the terms under which evolutionism was able to establish itself in the late

Victorian mind. There was no straightforward conflict between science and religion; evolutionism was eventually absorbed into the thinking of liberals and conservatives alike, although each side created a model of evolutionary development that suited its own tastes.

Theistic Evolutionism

The opposition was not slow to emerge. On one of his infrequent visits to the British Museum, Darwin was pointed out as 'the most dangerous man in England' by a clergyman.[3] Bishop Samuel Wilberforce's attack on the theory at the Oxford meeting of the British Association in 1860 is widely cited as a classic example of the confrontation with religion – and we should remember that Huxley's response is now known to have been much less effective than was once supposed.[4] By then the hostile reviews had already begun to appear in the periodical press and Darwin was forced to begin his long campaign to defend the theory.[5] He knew that many naturalists would find it difficult to accept natural selection but hoped that the general case for evolutionism would be reassessed. On the question of the theory's broader implications he still hoped that it would prove possible to reconcile it with a general belief that nature was given the power to develop by a benevolent Creator, but he knew that his own ideas on how the Deity's intentions were to be realized did not coincide with those of his contemporaries. As the criticisms piled up, it became increasingly obvious that there were many who found natural selection unacceptable as a hypothetical mode of divine control.

In his autobiography Darwin discussed at length his feelings on the religious implications of his theory. He admitted that at the time of writing the *Origin of Species* he was still a theist – that is, he still hoped to show that the universe was under divine control – but noted that over the following decades this belief had gradually weakened into agnosticism.[6] The Creator obviously did not design species individually, but Darwin thought that natural selection could only work if, on the whole, it tended to generate animals that could enjoy life in the environment to which they were exposed. This was compatible with the view that a benevolent God had instituted the evolutionary process as an indirect means of allowing species to remain adapted to an ever-changing world. Against this had to be set the undoubted suffering that existed in the world. The exponents of natural theology had always had to explain this away – for instance they had claimed that predators were the Creator's way of giving sick or aged animals a quick death. But natural selection turned struggle into a creative force. The unfit had to be eliminated if evolution was to advance, and suffering thus became a necessary feature of the world. Darwin could see the force of this objection and was thus compelled to accept the Creator as a somewhat remote figure who controlled the universe by general laws and could not be held responsible for the individual acts of suffering that were the by-products of the evolutionary process.

Darwin's continued belief in progress was almost certainly a consequence of his residual theism.[7] For a liberal thinker to see the universe as a purposeful system, change would have to have a meaningful direction. The universe had to do more than merely maintain a population of healthy living organisms – it must also show signs of advancing toward higher states of organization. Darwin's theory made it impossible to see the human race as the predestined goal of progress, since the metaphor of the evolutionary tree is incompatible with the belief that we stand at the head of a ladder of evolutionary progress. Pseudo-Darwinians such as Haeckel encouraged the belief that the tree of evolution did, after all, have a main stem aimed at the production of mankind, but Darwin's sense of progress was of a much less structured kind. He could only believe that each branch of evolution experienced an indirect tendency to move toward the production of more complex creatures. There would always be exceptions, as in the case of parasites which degenerate as they adapt to a less demanding lifestyle. Nevertheless, we have already seen how several passages in the *Origin of Species* were designed to ensure that the theory would be seen as providing indirect support for progressionism. In the 1860s and 1870s this was an important point in the theory's favour; only in later decades would a fascination with evolutionary degeneration emerge.

The one thing that Darwin could not admit was that God somehow played an active role in controlling the direction of evolution. This emerges clearly in the interaction with his leading American supporter, the botanist Asa Gray. In several of the essays reprinted in his *Darwiniana* of 1876, Gray used geographical evidence to good effect in defending the basic idea of adaptive evolution. But he was a deeply religious man who felt that he could only accept evolution if it could be seen as the unfolding of a divine purpose. At first he tried to argue that *any* mechanism of adaptive evolution is compatible with belief in God: better knowledge of the laws of physics does not change our faith in the existence of a God who instituted those laws, so why should better knowledge of the laws of biology pose any greater threat to belief? But in the end Gray found the element of random variation in Darwin's theory unacceptable. If God was to have any influence over the direction of evolution He must surely exert more positive control than this. Gray argued that species do not normally produce a host of purposeless variants needing to be eliminated – the 'scum of creation' as one critical reviewer had called them. 'Wherefore, so long as gradatory, orderly, and adapted forms in nature argue design, and at least while the physical cause of variation is utterly unknown and mysterious, we should advise Mr. Darwin to assume, in the philosophy of his hypothesis, that variation has been led along certain beneficial lines.'[8] In other words, God somehow ensures that the flow of new variations steers the evolution of each species in the desired direction.

Here, from one of Darwin's own supporters, was one of the most basic arguments against natural selection. The cause of variation was unknown, so why should we not assume that it actively directs evolution along purposeful channels? Darwin realized that this line of thinking would ultimately lead to natural selection being rejected as superfluous and he responded to Gray in a number of letters and then more openly in the conclusion to *The Variation of Animals and Plants under Domestication*. Here he compared natural selection to a builder who worked by picking out usefully shaped stones from those which had fallen from a cliff. In these circumstances no one would think that the stones fell from the cliff in accordance with the builder's desires. Even if we do not know the cause of variation the breeders had shown that random variants are always being produced. Hence if directed change was to occur, it was necessary to suppose that selection weeded out all the useless forms.

> However much we may wish it, we can hardly follow Professor Asa
> Gray in his belief "That variation has been led along certain beneficial
> lines" ... If we assume that each particular variation was from the
> beginning of all time preordained, then that plasticity of organization,
> which leads to many injurious deviations of structure, as well as the
> redundant power of reproduction which inevitably leads to a struggle
> for existence, and, as a consequence, to the natural selection or survival
> of the fittest, must appear to us superfluous laws of nature. On the other
> hand, an omnipotent and omniscient Creator ordains everything and
> foresees everything. Thus we are brought face to face with a difficulty
> as insoluble as that of free will and predestination.[9]

Darwin's lack of certainty in the face of a philosophical issue is typical,
but his objection to the idea that God somehow intervenes in the
evolutionary process is obvious enough. The laws grind on inexorably,
even if some of the consequences do not seem very pleasant to us.

For Darwin's opponents, however, the possibility that variation
might be directed along purposeful channels became the foundation
upon which they hoped to construct an alternative theory of evolution.
There were, of course, some efforts to retain an explicitly creationist
view of the origin of species. This was no doubt what lay behind
Wilberforce's attack at the Oxford meeting, and we know that Adam
Sedgwick was deeply upset by Darwin's renewed effort to replace divine
creation with natural law.[10] But by the late 1860s outright opposition
to evolutionism was becoming increasingly rare. Those who considered
themselves opponents of Darwinism were prepared to accept evolution
but not the claim that the process was governed solely by natural law.[11]
They objected to the haphazard open-ended aspect of development
implied by Darwin's theory, the assumption that, because variation is
essentially random, each branch of evolution is undirected except by
the pressure of the local environment. Someone told Darwin that the
astronomer Sir J. F. W. Herschel described the *Origin of Species* as 'the
law of higgledy-piggledy', and in his *Physical Geography* of 1861
Herschel wrote:

> An intelligence, guided by a purpose, must be continually in action to
> bias the direction of the steps of change – to regulate their amount – to
> limit their divergence – and to continue them in a definite course ... On

the other hand, we do not mean to deny that such intelligence may act according to law (that is to say on a preconceived and definite plan).[12]

The claim that God's plan might unfold through a process of development, widely rejected as heretical when proposed in Chambers' *Vestiges*, was now to become the foundation for the conservatives' opposition to the far more fundamental threat of Darwinian materialism.

The exponents of what has sometimes been called 'theistic evolutionism' believed that variation was an active force driving the species in a predetermined direction. Gray had suggested that adaptive evolution takes place because God has somehow influenced variation to supply the 'fittest' characters for the environment, and we shall see that Lamarck's theory of the inheritance of acquired characters was later called in to fulfil the same function. But many anti-Darwinians were convinced that evolution is *not* solely a function of adaptation to an ever-changing local environment. They thought that built-in forces must drive variation along regular paths that did not necessarily correspond to short-term adaptive requirements. They believed that the pattern displayed by the arrangement of living and fossil species was not an irregularly branching tree but an orderly scheme that gave clear evidence of underlying divine guidance.

In some respects the opponents of selectionism were more clearly aware of the implications that lay behind the theory than some pseudo-Darwinians. After all, Huxley and Gray both accepted directed variation, while Haeckel's progressionism implied that the ascent of life had a main line aimed at the production of mankind. But, with the exception of Gray, most pseudo-Darwinians were at least prepared to accept that the forces directing evolution were natural rather than supernatural. They might invoke a generally progressive force in evolution but they were unwilling to concede that development followed a pattern so regular that it could only be explained by a supernatural guiding force. The theistic evolutionists actively wanted to retain a role for the supernatural because this would allow them to defend the claim that social authority should still rest with religious leaders who alone could interpret the moral significance of the Creator's activity. Theistic evolutionism was thus the response of conservative thinkers anxious to resist the liberals' claim that progress occurred solely through the cumulative efforts of individual organisms (including human beings) struggling to cope with their environment.

The underlying issues were thus the same as those which had motivated the conservative opposition to Grant's Lamarckism and Chambers' *Vestiges*, only this time the liberal challenge was strong enough to force the conservatives into fighting a rearguard action to defend the concept of supernatural design *within* evolution. Darwin was vilified as he had feared back in the 1840s, but now there were enough thinkers who shared his own viewpoint to ensure that the opposition could not stamp out the new idea. In the 1830s and 1840s the anatomist Richard Owen had been one of the leading opponents of Lamarckism, and he had continued to defend the view that the history of life could only be understood as the unfolding of a divine plan. Darwin had, of course, had dealings with Owen from time to time, although he had been warned that Owen was a tricky character to deal with and the two men had never been close. The *Origin of Species* put Owen in a difficult position; he had not joined in the condemnation of *Vestiges* and had already hinted that the divine plan might unfold through non-miraculous causes, but he had been afraid to come out openly in favour of some form of evolutionism.[13] Now Darwin had reopened the general question of evolution but had done so by postulating a radical new mechanism that seemed to deny any direct control by the Creator. Owen has often been dismissed as an outright opponent of evolution but this is a gross oversimplification of his position. He certainly emerged as a leading opponent of *Darwinism*, but in fact he was trying to indicate his general agreement with the basic idea of evolution while repudiating natural selection as an expression of the materialism that he had opposed for decades.

Owen is thought to have primed Wilberforce for his attack at the Oxford meeting, but his own response came in a long review of the *Origin of Species* published in the *Edinburgh Review*.[14] Typically, Owen had written Darwin 'a most liberal note on the reception of my book, and said he was quite prepared to consider fairly and without prejudice my line of argument'.[15] But Darwin expected the response to be hostile and the attack in the *Edinburgh Review* fulfilled his expectations. The review was published anonymously, as was usual at this time, although there were few doubts about the authorship. The two-sidedness of

Owen's comments has often puzzled historians: on the one hand he criticized Darwin savagely, but on the other he often seemed to hint that there was nothing really new in the book. This apparent confusion was a product of his desire to oppose selection while leaving the door open for a form of evolutionism that would be compatible with his commitment to a divine plan of development. Darwin complained bitterly to Huxley about 'this false and malignant attack', which he thought had grossly misrepresented some of his arguments.[16] Owen maintained his opposition to the selection theory but a better indication of his real position can be seen in the conclusion to his *Anatomy of the Vertebrates* of 1868 in which he came out clearly in support of theistic evolutionism. He now proposed a theory of 'derivation' which 'sees among the effects of the innate tendency to change irrespective of altered circumstances, a manifestation of creative power in the variety and beauty of the results'.[17]

There were, in fact, a number of naturalists who shared Owen's feelings and who formed an obvious source from which an anti-Darwinian party might have been created. Unfortunately for them, however, their group lacked the tight cohesion of the Darwinians and they were unable to form a really effective opposition to the materialists' takeover of the sources of power within the scientific community. Owen was an influential figure, but he was a difficult person for even his fellow-travellers to get on with and he was largely out-manoeuvred by Huxley and Hooker. For this reason one of the most active centres of opposition came from outside the ranks of the biologists. The physicist William Thomson (later ennobled as Lord Kelvin) created a powerful, if indirect, argument against natural selection by attacking Lyell's uniformitarian geology.[18] In the late 1860s Kelvin argued that the earth cannot have been maintained in a steady state over vast periods of time because its interior is hot and must be gradually cooling down. Darwin had relied on the vast amounts of time allowed by Lyell's geology because he believed that natural selection was an immensely slow process. Kelvin's attack on uniformitarianism was, in effect, an indirect way of undermining the case for Darwinism, and Kelvin himself believed that evolution must have been speeded up by some divinely implanted progressive force. We now know that radioactivity supplies a source of energy that has kept the earth warm over vast periods of time but this was not realized until the early twentieth century. Darwin himself admitted that he had no answer to Kelvin's objections, although he refused to believe that Lyell could have been so much in error.

Kelvin's attack expressed the physicists' assumption that their subject established basic laws of nature that all other areas of science would have to accept. It also reflected the belief, widespread among physical scientists, that Darwin's theory did not meet the standards of scientific methodology. There was no experimental evidence that selection could produce new species (even Huxley conceded that this was a valid objection) and the theory made no testable predictions. Darwin, like all modern evolutionists, argued that the theory should not be judged by standards appropriate to a very different branch of science. Evolutionism's strength as a theory came from its ability to make sense out of a vast range of otherwise meaningless facts. This is a valid point but many physicists were unable to appreciate it – as are some of their modern descendants.

Kelvin's argument on the age of the earth was typical of the objections that would not go away and which led an increasing number of late nineteenth-century biologists to doubt the efficacy of natural selection. It is significant that Fleeming Jenkin's claim that natural selection would not work on the basis of blending heredity (discussed in chapter 8) was inspired partly by Kelvin's attack and was meant to illustrate the need for some other, more purposeful and hence more rapid mechanism of evolution. There was no shortage of objections aimed at making the same point, and the success of Darwinism rested in part on the skill with which Huxley and the other supporters managed to prevent the opposition coalescing into an effective force within scientific biology. Even so, there were a number of biologists who began to share the view that something more purposeful than random variation must direct the course of evolution. Like Owen, they looked for patterns in the fossil record and in the arrangements of living species, patterns that were too artificial to be the result of chance. At first the Darwinians were successful in dismissing their work as a return to old-fashioned idealism, but this way of thinking was never eradicated and staged a comeback during the later decades of the century.

A good illustration of the way in which the anti-Darwinian forces were out-manoeuvred, at least in the short term, can be seen in the fate of St George Jackson Mivart, who was virtually ostracized as a punishment for abandoning the Darwinian party. Mivart was a Catholic, but to begin with he had been quite willing to jump on the Darwinian bandwagon. In 1862 he obtained a Chair of Zoology at St Mary's College, London, with the support of both Huxley and Owen.[19] Within a few years, however, he began to express doubts that purely natural processes could explain the development of life towards mankind. He came to share Owen's belief in an underlying divine plan and began looking for anatomical and palaeontological evidence that would confirm the existence of evolutionary trends that could not be explained in purely naturalistic terms.

By the time he published his major attack on Darwinism, the *Genesis of Species* (1871), Mivart had already alienated the Darwinians. Darwin at first tried to minimize the antagonism: ' . . . I have just read (but not with sufficient care) Mivart's book, and I feel *absolutely certain* that he meant to be fair (but he was stimulated by theological fervour); yet I do not think he has been quite fair.'[20] Mivart himself attempted to stay on good terms, writing letters emphasizing his goodwill despite their differences of opinion. But a hostile review of the *Descent of Man* published by Mivart in the *Quarterly Review* convinced even Darwin that he was being less than honest.

> You never read such strong letters Mivart wrote to me about respect towards me, begging that I would call on him, etc., etc.; yet in the *Q. Review* he shows the greatest scorn and animosity towards me, and with uncommon cleverness says all that is most disagreeable. He makes me the most arrogant, odious beast that ever lived. I cannot understand him; I suppose that accursed religious bigotry is at the root of it.[21]

From this point on none of the Darwinians would have anything to do with Mivart. Huxley went out of his way to ridicule not only his science but also his attempts to show that Catholicism was compatible with a belief in evolutionism.

The *Genesis of Species* offered a cornucopia of anti-Darwinian claims which would be repeated (often without acknowledgement) by later generations of opponents right through to the modern creationists. Mivart emphasized the difficulty of understanding how complex organs could have been formed via a series of intermediate stages, each one of which must represent an adaptive improvement on the last. He argued that the intermediate stages between, say, a leg and a wing would be suited neither for walking nor flying. He also insisted that Darwinism offered no explanation of how the multitude of changes that must take place within the evolving organism could be coordinated. More positively he pointed to many relationships that seemed to imply something more than random variation as the source of evolutionary change. Why did vertebrates and cephalopods such as the octopus have very similar eyes? Surely this was too much of a coincidence and must indicate some divinely imposed predisposition for different lines of evolution to develop similar characteristics.

Mivart and Owen represented two foundations of anti-Darwinian thought – although Mivart's early support for Darwinism made it difficult for the two men to work together. There were plenty of biologists who shared their hostility to the Darwinians' claim that all could be explained in terms of natural law. W. B. Carpenter, a physiologist who at first supported Darwin, later argued that evolution displayed patterns too regular to be accounted for by natural selection. The Duke of Argyll's *The Reign of Law* invoked the beauty of species such as the hummingbirds as evidence that the Creator intended evolution to have more than purely utilitarian purposes. But theistic evolutionism was never a very effective line of opposition, partly because its supporters did not form a coherent school of thought and partly because the idea was too obviously a compromise with the old supernaturalism. The late 1860s and the 1870s represent the high point of Darwinism's influence, as Huxley and his supporters effectively marginalized the anti-Darwinians within the British scientific community.

In the later decades of the century, however, there was a resurgence of opposition that swept even pseudo-Darwinism aside. The arguments put forward by Mivart and Owen were never really answered, at least to the satisfaction of those who continued to believe that nature unfolded to reveal a divine purpose. The whole point of Mivart's attack was to show that evolution was a more orderly and more purposeful process than mere natural selection of random variation by the local environment. His arguments could thus be exploited by later naturalists who were no longer committed to the idea that the source of orderly development was the supernatural guidance of variation. Later generations of anti-Darwinians would argue that the guidance came from variation trends that were somehow controlled by the forces that shape the growth of the individual towards maturity. This freed the search for evolutionary regularities from the outright supernaturalism of Mivart's position and gave it a new lease of life in the later decades of the century. Some palaeontologists adopted the theory of 'orthogenesis', in which many aspects of evolution were seen not as adaptive responses to changes in the environment but as the unfolding of rigidly predetermined trends.[22] This theory was based on the assumption that variation, far from being random as Darwin had supposed, was forced in a particular direction by the genetic constitution of the species. In thus rejecting the adaptive basis of evolution in favour of predetermined trends, the supporters of theistic evolutionism and of orthogenesis were challenging the utilitarian foundation that lay at the heart of Darwinism. The fact that this line of thought grew rather than diminished in strength during the late nineteenth century should warn us that the success of the early Darwinians was only a temporary affair.

The Rise of Lamarckism

Even in those cases where evolution *was* adaptive, the opponents of Darwinism sought an alternative to natural selection that would seem more in tune with their belief that evolution expresses an underlying divine purpose. They found such an alternative in Lamarck's old mechanism of the inheritance of acquired characteristics. Darwin himself had never ruled out the inheritance of acquired characters as a supplement to natural selection, and in response to criticism he often insisted that the effect might help to explain cases where the selection theory seemed to run into difficulties. But to his opponents Lamarckism soon came to represent a major component in the search for an alternative that would uphold an image of the evolutionary process as a more orderly and purposeful process.

Lamarckism seemed to offer a more purposeful mechanism than natural selection because it allowed the behaviour of animals to play an active role in directing the course of evolution. Instead of killing off all the non-adaptive forms produced by random variation, Lamarckism assumed that variation was shaped by the individual's own efforts to adapt itself to its chosen lifestyle. Evolution was directed by the efforts made by successive generations of animals striving to cope with changes in their environment, the results of which were accumulated by inheritance. In the classic example, all members of the giraffe species began to stretch their necks to reach the leaves of trees. Extra neck length comes about because the members of the species actively seek to adapt themselves to the new way of feeding, and their individual efforts accumulate over many generations to produce a significant effect on the species. The new feeding habit directs the species' evolution along a course leading to increased specialization. The inheritance of acquired characters thus provides a *natural* force directing variation along useful channels. Although once rejected as the epitome of materialism, Lamarckism could now be seen as a more purposeful alternative to natural selection – much more the kind of mechanism that would have been instituted by a wise and benevolent God. The fact that Darwin himself accepted an element of Lamarckism shows that in the decades before the emergence of Mendelian genetics this was a perfectly plausible hypothesis about how evolution might take place.

The orthodox history of modern evolutionism tends to dismiss Lamarckism and other non-Darwinian mechanisms as being of minor interest. The fact that Darwin himself could not rid himself of a residual belief in the inheritance of acquired characters is presented as a rather embarrassing sign of his failure to anticipate the need for a theory of fixed hereditary units. The main line of development leads from Darwin's initial formulation of natural selection to the modern synthesis of selectionism and genetics. But a closer look at the evolutionism of the late nineteenth century reveals that Lamarckism played a far more fundamental role. The theistic evolutionism of Owen and Mivart gradually gave way to a more naturalistic anti-Darwinism based on Lamarckism and orthogenesis, that is, on evolution by adaptive and non-adaptive modifications of individual growth. The role of Lamarckism has been obscured because in the British situation it is

associated with Samuel Butler, a vitriolic opponent of Darwinism who
was ignored by the scientific community. But Butler was by no
means the only late nineteenth-century Lamarckian, and his hostility
to Darwin should not be allowed to blind us to the fact that his
views represent a major current of evolutionary thought.

Lamarckism is normally associated with conservatives who oppose
the 'materialism' of the selection theory. It is certainly true that the
hypothesis has been supported by a long tradition of anti-materialists
stretching from Samuel Butler through George Bernard Shaw to Arthur
Koestler.[23] But even among the 'Darwinians' of the 1860s there were
some who gave the inheritance of acquired characters a much more
significant role. Far from being a product solely of an idealistic anti-
Darwinism, Lamarckism could be taken up by the most fervent supporters of the liberal progressionism that took Darwin as its figurehead.
Huxley, it is true, never had any time for the theory, but the social
philosopher Herbert Spencer – next to Darwin and Huxley the most
influential evolutionist of the Victorian era – always used it as one of
the principal foundations of his thought. Spencer had, in fact, come
out openly in support of transmutation long before the *Origin of Species*
was published. His essay 'The Development Hypothesis' of 1851[24] had
tried to promote Lamarckism as the basis for a new assault on the
question of organic origins. Spencer had been unable to interest
working scientists in this long–discredited possibility, but once the
Origin of Species had broken the ice, he was able to emerge as the
philosopher of the evolutionary movement by adopting natural selection and Lamarckism as the twin pillars of a unified theory of adaptive,
progressive evolution based on the cumulative effects of the day-to-day
activities of living organisms.

Spencer's philosophical position was a perfect expression of the
liberalism to which Darwin and Huxley both owed allegiance. He was
the apostle of *laissez-faire* individualism, determined to argue that social
progress would follow inevitably once archaic restrictions upon individual freedom were abolished. Biological evolutionism was an essential
foundation for his cosmic progressionism – a guarantee that human
progress was merely an inevitable continuation of a natural trend. It
was thus important for Spencer, like Darwin, to show that biological
evolution resulted not from some divinely implanted force directing

variation but from the normal activities of living things going about their daily business. Small wonder, then, that Spencer was accepted into the X Club and became a member of the inner circle governing the liberal takeover of the scientific community. Darwin himself always admired Spencer's genius, although he wished he would devote more effort to observation and less to abstract speculation: 'I feel rather mean when I read him: I could bear, and rather enjoy feeling that he was twice as ingenious and clever as myself, but when I feel that he is about a dozen times my superior, even in the master art of wriggling, I feel aggrieved. If he had trained himself to observe more, even at the expense, by the law of balancement, of some loss of thinking power, he would have been a wonderful man.'[25]

Spencer believed that evolution was driven by individual competition and he was thus quite willing to accept the idea of natural selection once it was pointed out to him. Indeed, it was he who coined the term 'survival of the fittest' – a phrase which Darwin welcomed and which has become a potent symbol of the Darwinian emphasis on struggle.[26] Spencer's view that social progress would only come about if all human activity was governed by the principle of free enterprise has led to his being singled out as a typical 'social Darwinist' – a clear indication of the congruence between Darwin's thinking and the competitive ethos of Victorian capitalism. Yet Spencer did not abandon Lamarckism for natural selection: his *Principles of Biology* urged that both the selection of random variation and the preservation of acquired characters must cooperate in the evolutionary process. Struggle was important not just because it weeded out the unfit but because it encouraged all organisms (including human beings) to become fitter in order to escape the suffering that was the consequence of failure. Their efforts to cope with changes in their environment would produce useful new characters (such as the giraffe's longer neck) which might then be passed on to future generations. Without struggle there would be no pressure to make these personal adaptations. In Spencer's eyes, Lamarckism and natural selection both expressed possible ways in which the effects of competitive activity could accumulate to give social or biological evolution.

In some respects Spencer's emphasis on the individual's struggle to better itself expressed the Victorians' faith in self-help even more clearly than Darwin's natural selection. The problem with natural selection, as far as Spencer was concerned, was that it allowed the individual no freedom to improve itself by its own efforts: if it was born with a harmful character then it was doomed. Only by allowing for individual self-improvement to participate in evolution was it possible to capture the spirit of the free-enterprise system completely. Much of what has been called 'social Darwinism' is thus really social *Lamarckism* based on Spencer's highly influential ideology of progress through struggle. Natural selection mirrored only the negative side of this social philosophy, the elimination of those unfortunates who had no ability to change in the face of environmental stress.

Along with natural selection, Lamarckism thus provided an important ingredient of the liberal ideology lying at the heart of Darwinism's original appeal. Far from eliminating the older hypothesis, Darwin's precipitation of the general conversion to evolutionism gave it a new lease of life. In the liberal ideology, the important aspect of the inheritance of acquired characters was its implication that the individual had the power to adapt itself to a new environment. Since Spencer believed that the inevitable consequence of this process over many generations would be progressive evolution, he shared Darwin's early faith in the purposefulness of nature's activity. Although Spencer himself had no time for religion his views were taken up by liberal Protestants as a modernized version of the traditional work-ethic: God had delegated to nature the power to reward those whose efforts contribute toward the future perfection of the race.[27] The claim that progress arose inevitably, if indirectly, out of the activities of individuals could be treated as a way of evading the apparently atheistic implications of Darwinism. The liberals could be confident that what they were doing was contributing to the moral development of the universe; and those who wished to retain some elements of traditional Christianity could console themselves by arguing that, for all its apparent harshness, nature really did express a divine purpose.

It is significant that Spencer never abandoned his support for Lamarckism. When the German biologist August Weismann began to argue for a 'neo-Darwinism' purged of the Lamarckian element during the 1880s, Spencer wrote openly in defence of the inheritance of acquired characters.[28] The claim that heredity rigidly determined the individual's character was never a part of his philosophy. When Weis-mann (quite correctly) pointed out that natural selection threw all the emphasis onto such rigidly determined characters, Spencer faced a crisis of loyalties. If 'Darwinism' was now to be redefined in a way that eliminated all the non-Darwinian mechanisms that even Darwin himself had permitted, Spencer would have to speak out against what he saw as a dogmatic and oversimplified selectionism. Having begun as an important member of the Darwinian camp, Spencer was thus forced into opposition once the more flexible kind of pseudo-Darwinism began to break down. His change of attitude quite clearly illustrates the extent to which the Darwinism of the 1860s and 1870s had incorporated elements that would not survive into the twentieth century.

Conservatives such as Owen and Mivart opposed the liberal ideology that lay behind the original version of Darwinism. To defend their belief that human nature is something more than a product of untold generations of selfishness, they wanted to show that evolution displayed more direct evidence of God's designing hand. They did not believe that individual activity alone directed evolution: some higher Power must be seen to be at work guiding nature towards higher things. For this reason they repudiated Darwinism's reliance on adaptation as the sole driving force of progress. And yet the conservatives were forced to accept that some evolution was adaptive, and in their eyes the element of purposefulness in the Lamarckian mechanism made it far more acceptable than natural selection as an expression of the Creator's handiwork. In the end the link between Lamarckism and the liberal ideology was broken and a new link forged with the belief that nature expresses a divine purpose manifesting itself in the evolutionary process. One of the earliest exponents of this interpretation was the novelist Samuel Butler who – like Mivart – found it necessary to insult Darwin personally and paid the penalty of being ostracized by the scientific community.[29]

Butler was a young man working on a sheep farm in New Zealand when the *Origin of Species* appeared. He was an instant enthusiast for the new idea and wrote letters to Darwin proclaiming his support. In 1872 his novel *Erewhon* included what many took to be a satire on natural selection, since the Erewhonians have destroyed all their machines in case they become too efficient and displace the human race itself. Butler again wrote protesting that this was not his intention and actually visited Darwin at Down on several occasions. He had now become interested in the way in which instincts are formed, which he supposed to occur when long-standing habits eventually become fixed by heredity. This was a very Lamarckian notion, and in the course of writing his book *Life and Habit* Butler began to realize that he was now moving towards a non-Darwinian theory of evolution. The crisis seems to have come through a reading of Mivart's *Genesis of Species*, which convinced Butler that it was possible to be an evolutionist without being a Darwinist. Mivart mentioned Spencer's Lamarckism, and Butler now went back to read Lamarck himself along with other early evolutionists including Buffon and Erasmus Darwin. He soon became convinced not only that the inheritance of acquired characters was superior to natural selection but that Darwin had been less than honest in his failure to acknowledge the role of his predecessors.

In his *Evolution Old and New* of 1879 Butler developed the claim that the inheritance of acquired characters offered a way of reconciling evolutionism with natural theology. The natural selection of random variation was a process of trial and error incompatible with any belief that a Creator sustains the universe. But if animals could control their own evolution through choosing new habits that would adapt them to any changes in their environment, it would be possible to see their purposeful behaviour as an expression of a creative Mind working within nature. Instead of designing species from outside, so to speak, God works through the purposeful behaviour of the animals to which He has delegated His creative power. Butler later wrote privately to Mivart to argue his case that the best way of preserving a role for the Creator was to internalize Him within the evolutionary process.[30] Although Butler shared Spencer's emphasis on individual modifications as the basis of evolution, he gave Lamarckism a new dimension by emphasizing the role of creative choice in the determination of new habits.

Just before *Evolution Old and New* was published, an article on Erasmus Darwin by Dr E. Krause appeared in the German periodical *Kosmos*. This was subsequently enlarged and translated as part two of a *Life of Erasmus Darwin* that Charles himself had prepared. Darwin sent Krause a copy of Butler's book, and some disparaging remarks about it were incorporated into the English version of Krause's text. Unfortunately Darwin forgot to explain in his preface that the translation was an extended version of the original German text. Butler checked the original and found that there was no reference to his book. He became convinced that the translation was an underhanded way of trying to show that a German scholar had already come out against his views even before the publication of *Evolution Old and New*. Butler wrote to Darwin with his complaint and received an apology, but he was not satisfied and accused Darwin publicly of misconduct in a letter to the *Athenaeum*. Darwin wanted to reply in the same forum and was encouraged in this by his son Francis, but the rest of the family, together with Huxley, argued that it would be beneath his dignity to get involved in such an exchange. Butler of course interpreted silence as an admission of guilt and as further evidence that he was being snubbed by the Darwinians. He wrote a number of later books condemning natural selection in the strongest terms. In an 1890 response to Weismann's neo-Darwinism he wrote: 'To state this doctrine is to arouse instinctive loathing; it is my fortunate task to maintain that such a nightmare of waste and death is as baseless as it is repulsive.'[31]

The original Darwinians tried to brush Butler off as though he were merely a gadfly, and this convinced him that scientific biology had been taken over by a clique determined to impose selectionism as a rigid dogma. By the time he wrote the response to Weismann quoted above, however, Butler was able to report that the tables were being turned. Far from arousing mass support, Weismann's all-or-nothing selectionism was alienating even some of the original Darwinians such as Spencer. Many biologists now began to accept that Lamarckism did offer the key to a less materialistic evolutionism just as Butler had suggested.[32] Perhaps the clearest indication of this is the more favourable attitude that Francis Darwin now began to adopt toward Butler's views. As a respected biologist, Francis was in a position to provide the kind of scientific recognition that Butler had craved. On several occasions he spoke openly in favour of the analogy between heredity and memory that lay at the heart of Butler's Lamarckian explanation of instinct.[33] After Butler's death he collaborated with his biographer H. F. Jones in the production of a pamphlet giving a more balanced account of the whole affair.[34]

The renewed support for Lamarckism coupled with the emergence of the theory of orthogenesis suggest that the Darwinians of the 1860s and 1870s did not retain control of the scientific community. The movement had never been a perfect expression of Darwin's own views because its supporters had admitted an even wider range of non-Darwinian mechanisms than Darwin himself. But it had been held together by the liberal ideology promoted by Spencer and Huxley and had paid lip-service, at least, to the claim that natural selection was a major component of the evolutionary mechanism. After the death of Darwin in 1882 it fell apart as more and more biologists responded to Weismann's dogmatism by turning to less mechanistic explanations of development.

In America the emergence of an explicitly anti-Darwinian movement had begun even earlier. Asa Gray had helped to convert the American scientific community to evolutionism but the influence of the idealist Louis Agassiz, now teaching zoology at Harvard, ensured that many Americans would be unwilling to accept Darwinian naturalism. Agassiz's disciples provided the kind of organizational network that the British anti-Darwinians lacked. Palaeontologists such as Edward Drinker Cope and Alpheus Hyatt became the focus for a coherent school of Neo-Lamarckism from the late 1860s onwards.[35] Like Mivart and Butler, they wanted to see evolution as the unfolding of a divine plan and sought evidence for this in the regular patterns that they thought they could discern in the development of some groups within the fossil record. Beginning as theistic evolutionists, they moved on during the 1870s to lay the foundations for the late nineteenth-century explosion of interest in both Lamarckism and orthogenesis. The American palaeontologists realized that Lamarckism could itself become the basis for a theory in which the course of evolution was predictable. Once a population had made the choice of a new lifestyle its descendants could do little else but carry on reinforcing the line of specialization that had been mapped out for them. Hyatt also believed that, once a species had run out of evolutionary 'energy', it would degenerate in a more or less predictable manner to a senile form as a prelude to extinction. His theory thus combined Lamarckism with what would later be called orthogenesis.

The fact that an actively anti-Darwinian school of thought could emerge in America at so early a date makes it clear that the Darwinian takeover of the British scientific community was the product not of the overwhelming success of Darwin's arguments but of clever man-oeuvring by Huxley and the other liberals to isolate and divide the conservative opposition. We have seen that they were successful in the short term, but by the last decades of the century the liberal ideology was on the wane in the face of growing support for imperialism. In the new climate of opinion the organizational support for Darwinism fell apart and all the old anti-Darwinian arguments emerged to plague the theory once again. What happened in America during the 1860s and 1870s was mirrored a decade or two later in Britain as the supporters of Lamarckism and orthogenesis proclaimed themselves to be outright opponents of Darwinism.

The traditional interpretation of Darwin's influence tends to play down the effectiveness of the opposition in order to celebrate the triumph of Huxley, Spencer and their liberal ideology of naturalism. The modern resurgence of creationism also tends to make us forget that there have been anti-Darwinians who were nevertheless still evolutionists. It is all too easy to lose sight of the fact that the initial triumph of Darwinism is separated from the theory's modern success by an interlude in which anti-Darwinian thinking was common even within scientific evolutionism. Only by carrying the story into this intermediate period can we get the breadth of perspective that allows us to appreciate the very special circumstances that allowed Darwin's book to become the figurehead for the conversion of the scientific world to evolutionism. The original Darwinian debate was a close run thing because even pseudo-Darwinism (to say nothing of Darwin's own theory) was not a self-evident solution to the scientific and conceptual problems of the age. Darwin succeeded because his theory could be taken over and exploited by the exponents of a particular ideology who were, at least in the British context, successful in using the *Origin of Species* as the spearhead for their assault on the bastions of conservative thought. But the opposition was stalemated, not destroyed, and returned with renewed vigour as the century progressed towards its close.

10

Human Origins

<hr/>

T HE BROADER IMPLICATIONS of the debate over the mechanism of evolution were brought more sharply into focus by the theory's application to the origins of humankind. Many people were upset by the very idea that we might be descended from apes. It was this issue which led to the uproar at the Oxford meeting of the British Association, when Wilberforce asked Huxley if he claimed to be descended from an ape on his mother or his father's side – and Huxley replied in similarly blunt terms.[1] In 1864 Benjamin Disraeli was invited by Wilberforce to speak against materialism in the Sheldonian Theatre at Oxford and made his famous declaration: 'The question is this – Is man an ape or an angel? My Lord, I am on the side of the angels.'[2] A host of cartoons in the popular press lampooned the notion that we might be related to the gorilla. By raising the possibility of such a link the evolutionists were threatening the concept of the immortal soul and hence the traditional foundations of morality. If evolutionism were true, a new source of moral values would have to be found – and where was this to come from if the mechanism of evolution which produced us was (in Butler's words) a nightmare of waste and death? Could the products of such a process be expected to behave with more nobility than the brutes themselves?

PLATE 14 Caricature of Darwin linking his views on human origins to his work on earthworms, from *Punch*. (Reproduced by kind permission of Punch publications)

Darwin himself had faced up to these problems at an early stage in the formulation of his theory. The M and N Notebooks from the late 1830s had already laid the foundations for his assault on the traditional view of the human mind and its moral capacities. From the start he accepted that we were nothing more than highly developed animals and sought to explain our social behaviour in biological terms. Yet he knew that these issues lay at the heart of the conventional mistrust of evolutionism. Even a middle class liberal could be ostracized for daring to undermine the fabric of society in this way. The reaction to Chambers' *Vestiges* revealed how sensitive the issue was and confirmed Darwin's feeling that it was necessary to deflect attention to more technical questions so that the case for evolutionism could be discussed more freely. The *Origin of Species* thus avoids the question of human origins except for a single statement that Darwin felt honour-bound to include so that he could not be accused of concealing his beliefs: 'Light will be thrown on the origin of man and his history.'[3] Even without this statement the subject was bound to come to the fore, and the debate raged throughout the 1860s before being revived by Darwin's own contribution, the *Descent of Man*, in 1871.

Our discussion of the issues must once again balance Darwin's own contributions against the public response to his and other evolutionists' ideas. There can be no doubt that Darwin made a bold effort to explore the implications of his theory for mankind, pioneering a range of studies designed to reveal the animal origins of human faculties. At the same time, however, it is necessary to recognize that he shared some of the preconceptions about human nature that were common among Victorian Englishmen. In particular he believed that white males represented the most highly developed form of humanity. These preconceptions are important because they shaped the public response and determined the circumstances under which the basic idea of an animal origin for mankind would eventually become acceptable. We must also beware of distortions that can all too easily be introduced into our interpretation of the debate by focusing too strongly on the positions adopted by extremists. The basic evolutionary position was indeed adopted by a majority of late Victorian thinkers, but their beliefs about *how* we emerged from the apes did not necessarily follow Darwin's own suggestions and certainly did not anticipate the modern viewpoint.

Once again the idea of purposeful development became crucial: people found that they could reconcile themselves to the prospect of an animal ancestry provided that the evolutionary process was seen as a force driving nature towards a morally significant goal. Instead of seeing ourselves as standing above nature by virtue of our possession of an immortal soul, we became the cutting edge of nature's drive toward the generation of ever-higher mental states.

This interpretation of Darwinism's impact challenges once again the claim that evolutionism forced the Victorians to adopt an openly materialistic philosophy. For Butler and the opponents of the selection theory the 'survival of the fittest' epitomized a brutalizing image of nature which threatened to deny mankind any hope of aspiring to higher things. There could be no meaningful role for mind or morality in such a system. Many of Darwinism's modern opponents continue to stress its materialistic implications and are thus encouraged to assume that those implications were forced upon the Victorians by their accept-ance of the theory. It is often claimed that rampant capitalism and imperialism encouraged the adoption of a ruthless policy of 'social Darwinism' in which unfit individuals and unfit races were condemned to death or slavery in the name of progress. Brute force replaced the traditional virtues of love and sympathy thanks to the influence of Darwin's teaching.

Historians now realize that such an interpretation overdramatizes the impact of evolutionism because it ignores the efforts that were made to see a moral purpose in the evolutionary process itself. Even many of the so-called 'Darwinists' failed to appreciate what we now regard as the obvious implications of the selection theory. For them, nature may have been harsh – but it was not a meaningless chaos because the harshness was designed to promote the advance of life toward higher levels of consciousness. Liberals who sought to undermine the tra-ditional view of morality, with its strong links to a conservative view of social relations, were themselves convinced that social evolution constituted an inevitable progress towards a morally significant goal. Spencer's evolutionary philosophy saw progress as the automatic con-sequence of the cumulative efforts of generations of individuals strug-gling to cope with their environment. Anthropologists too saw modern European civilization as the highest rung on a ladder of social progress that all races could eventually hope to climb. Far from being a product of Darwinian materialism, the theory of social evolution emerged independently of its biological equivalent and helped to promote a progressionist viewpoint that subverted the logic of Darwin's theory of branching development.

The Descent of Man

A number of factors in Darwin's early career helped to pave the way for his later study of human origins. He came from a family that was deeply involved in the campaign against slavery and we have seen that he was revolted by the treatment of the black slaves he witnessed in South America. He was thus encouraged to see all the races of mankind as members of a single species, unlike the 'polygenists' who dismissed the blacks as separately created species designed by God to be sub-servient to the whites. His experiences with the natives of Tierra del Fuego, however, had given him an impression of just how bestial the lives of the 'lowest' members of the human race could be. Once Darwin had freed himself from the idea of divine creation, the Fuegians' lifestyle as primitive hunter–gatherers must have seemed a perfect illustration of the conditions under which the earliest members of the human race had lived before the emergence of agriculture or any form of civilization. Many later evolutionists would assume that such primitives were 'living fossils' – relics of the ancestral form of mankind still surviving in out of the way parts of the world. But Darwin knew from experience that this was an oversimplification. The Fuegians might have no concept of God and no morals (it was claimed that they ate their old women in times of hardship) but the three whom Fitzroy had taken to England had been educated into some semblance of civilization. Darwin realized that the Fuegians' way of life had been forced on them by the bleak conditions to which they had been compelled to adapt.

If the Fuegians were truly human, not a 'missing link' between apes and men, their lack of civilized values gave a clue to the mental processes by which such values are imprinted on the minds of those raised under more fortunate conditions. Clearly the religious and moral senses were not divinely implanted instincts – they were habits of thought created in our minds by education. Darwin would have been aware of the long tradition of 'sensationalist' philosophy which sought to explain the working of the mind solely in terms of how sense impressions are related to give an image of the world in which we live.[4] Erasmus Darwin had claimed that even animals have no instinctive behaviour patterns imprinted on their brains; all their behaviour is shaped by an intelligent response to the environment. This position was bitterly criticized by naturalists who claimed that each animal species has been endowed by its Creator with instincts appropriate for its way of life. Such a viewpoint fitted in neatly with the belief that the human con-science represents a divinely created moral instinct. It was Lamarck who sketched in the outlines of an evolutionary compromise that would deeply influence the young Darwin. Lamarck suggested that instincts are, in effect, learned habits that have been followed for so long that they have been converted into hereditary instincts. The inheritance of acquired characters works for mental functions as well as physical structures such as the giraffe's neck. Evolution requires the conversion of learned mental habits into biologically imprinted behaviour patterns; in effect, instinct is unconscious memory inherited from previous generations.

This Lamarckian position was to become the basis for much nineteenth-century thought on the origins of human faculties. If evolution worked in this way, then intelligence was the guiding force of behaviour at all stages of development and one could expect intelligence to be increased through time as each generation struggled to cope with its environment. At the same time, successful behaviour patterns would be converted to instincts, leaving the intelligence free to deal with the fresh challenges that confront each new generation. Even the higher faculties of mankind, including our much vaunted moral powers, would originate as instincts developed by evolution to adapt us to living in social groups. As our ancestors learned to live in tribes, so they would have built up the instincts that would regulate behaviour to make cooperation and other forms of social interaction possible. Such a position certainly destroys the claim that the conscience is a divinely implanted moral instinct, but it does not turn us into the blind automata of the anti-materialists' nightmare. In their origins, at least, the moral powers have been purposefully created by the cumulative mental activity of earlier generations. Mind has not been banished from nature; it has been incorporated into it as the guiding force of evolution.

It was along these lines that Darwin himself began to think as he explored the human implications of evolutionism in his M and N Notebooks. He was convinced that all instincts, including the social instincts of mankind that we dignify by the name of morality, have been created by evolution. In effect, he was trying to turn morality into a branch of biology through the proposal that our instinctive behaviour can only be understood as a product of the natural processes that have adapted us to a particular way of life based on the family unit as a means of raising children. The social (or moral) instincts have been implanted because they are useful, not because they represent some higher power imposed on us from without. However, Darwin was also acutely aware of the personal implications of what he was doing. He studied his own children as a means of trying to understand the order in which the various mental functions appear in the developing mind. This, he felt, might throw some light on the process by which those functions had been created by evolution. It was certainly not his intention to claim that people are mindless automata and at this stage he still felt that the overall nature of the evolutionary process was the Creator's way of generating higher mental functions in the world.

Unlike the vast majority of his fellow evolutionists, Darwin gradually came to realize that the Lamarckian mechanism that converted habits into instincts was not the only way of explaining how evolution shaped the creation of new behaviour patterns. This process cannot explain the instincts of neuter insects such as ants and bees, since these by definition cannot transmit any newly acquired habits to their offspring. In the course of the 1840s and 1850s Darwin gradually came to realize that natural selection could itself explain how instincts are changed to meet a changing lifestyle. On the assumption that instincts are somehow imprinted into the physical structure of the brain, they must be subject to some individual variation just like any other character. Natural selection can then pick out those individuals whose instincts have varied in a favourable direction. This offers a far more materialistic view of evolution because it no longer allows the intelligent actions of earlier generations to direct the creation of new instincts. Mental evolution, like physical, becomes a process of trial and error based on the selection of random variation. It was this aspect of Darwin's later views that particularly offended Lamarckians such as Samuel Butler.

The application of the selection theory to the origin of animal instincts, especially in difficult cases such as neuter insects, forms a substantial chapter of the 'Natural Selection' manuscript that Darwin began to prepare in the late 1850s, and, of course, a chapter of the *Origin of Species*.[5] Much of Darwin's work on the subject was subsequently published by Romanes, who became his disciple in the field of mental evolution (see below). But the public response to the *Origin of Species* persuaded Darwin that he must address the question of human origins more directly. It was here that he had to convince his opponents (and even some of his supporters) that evolutionism could bridge what had hitherto been accepted as a fundamental gulf between the animal and human levels of mentality. In the end he would have to invoke both natural selection and Lamarckism in order to make his argument as flexible as possible.

The argument began over the fairly technical question of how closely the human species was related to the great apes. Lamarck had made it clear that mankind must have evolved from something like an ape and this hypothetical ancestry was already closely associated with evolutionism in the public mind. Darwin himself certainly felt that we must have evolved from a creature that, were it alive today, we should have to classify with the apes. The popular impression was that we have evolved from a modern ape such as the gorilla, but Darwin's branching model of evolution made it quite clear to him that this was not the case. The ape and human branches must have diverged from a common ancestor, but none of the living apes would have preserved that ancestral form completely. In the absence of fossil evidence one could only assume that the common ancestor would have been an ape with a less specialized structure than any of the living apes.

Opponents of evolutionism leapt upon the general idea of an ape ancestry and attempted to discredit it by claiming that there was, in fact, a considerable gulf between mankind and the living apes, a gulf which (they supposed) it was impossible for nature to bridge. For all that he was inclined to the general idea of natural development, Richard Owen had been campaigning against the link between humans and apes since his confrontation with the radical Lamarckians in the 1840s.[6] In 1858 he returned to the subject and declared that there was an immense gulf between the anatomy of the ape and human brains, especially the lack of the organ known as the hippocampus minor in the ape brain. He repeated this position in a session at the 1860 British Association meeting a few days before the famous Huxley–Wilberforce confrontation. Huxley had already indicated his opposition to Owen's views even before the *Origin of Species* was published, and at the Oxford meeting he flatly contradicted Owen, promising to substantiate his claims in print as soon as possible. In an article in 1861 Huxley demonstrated that the differences between the brains of humans and apes were of degree only – there are no distinctively human structures in the brain that could be responsible for our higher faculties. His *Man's Place in Nature* of 1863 expanded on this theme at length, showing that humans are more closely related to the apes than the apes are to the monkeys.

Huxley had demonstrated that there was no physical gulf between humans and apes but he had not proved that the human race had an ape ancestry, nor had he shown how the higher mental faculties could have evolved from the apes' level of mental activity. This latter point was the most crucial, and even some of Darwin's closest supporters found themselves in agreement with the objections raised by more conventional thinkers. Lyell had long harboured doubts about evolutionism precisely because it appeared to threaten the spiritual status of mankind by linking us to the brutes. In his *Antiquity of Man* of 1863 he suggested that if evolutionism were true the human race must have appeared by a sudden leap or saltation, thus ensuring that we were not too closely linked to the apes. Darwin had always been anxious to avoid making this kind of an exception for mankind, and he complained that this part of Lyell's book made him groan.[7]

Equally dangerous was the defection of Wallace, who became converted to spiritualism in the course of the 1860s and was thus led to doubt that natural selection could ever have produced the higher qualities of the human mind. In a review in 1869 of the tenth edition of Lyell's *Principles of Geology*, he openly suggested that some higher power must have steered human evolution in the appropriate direction, a view developed at length in an article included in his *Contributions to the Theory of Natural Selection* of 1870. Darwin wrote: 'As you expected, I differ grievously from you, and I am very sorry for it. I can see no necessity for calling in an additional and proximate cause in regard to man.'[8] Wallace's acceptance of the traditional view that the human mind stands above nature illustrated to Darwin the necessity for him to put his own views before the public. He was already working on the *Descent of Man*, which appeared in February 1871. Darwin complained to Hooker; 'I finished the last proofs of my book a few days ago; the work half killed me, and I have not the most remote idea whether the book is worth publishing.'[9] In fact the *Descent of Man* was to prove a major contribution to the debate over human origins.

Darwin was able to rely on Huxley and other authorities to support the view that the human race had strong anatomical resemblances to the apes. But the mental functions were his chief concern, and much of the *Descent of Man* was taken up with evidence designed to convince the reader that the higher faculties were not unique to mankind. Darwin cited numerous examples of animal behaviour that seemed to indicate that dogs, apes and other higher animals possessed at least rudimentary elements of intelligence and even of the moral sense.

> Several years ago a keeper at the Zoological Gardens shewed me some deep and scarcely healed wounds on the nape of his own neck, inflicted on him, whilst kneeling on the floor, by a fierce baboon. The little American monkey, who was a warm friend of this keeper, lived in the same compartment, and was dreadfully afraid of the great baboon. Nevertheless, as soon as he saw his friend in peril, he rushed to the rescue, and by screams and bites so distracted the baboon that the man was able to escape, after, as the surgeon thought, running great risk of his life.[10]

Many of these cases can now be seen to rest on an anthropomorphic interpretation of animal behaviour that would be repudiated by modern scientists. But it was important for Darwin to create the impression that all the human faculties had some origin in the lower animals. Mental evolution would thus consist of an increase in the level of these faculties, not the creation of something entirely new.

Darwin devoted another book to making the opposite point that human behaviour shows many relics of our animal ancestry. His *Expression of the Emotions in Man and the Animals* was published one year later (1872) and was intended to demonstrate that our emotional behaviour follows patterns that are already visible in the lower animals. The curling of the lips into a sneer may be a relic of the snarling action designed to show the teeth to an enemy when the teeth were still used as weapons. By means of such examples Darwin attempted to convince his readers that our behaviour is not lifted so far above that of the lower animals as we normally imagine. Our lives are still dominated by functions imposed on us as a result of our animal ancestry.

It was the *Descent of Man*, however, which tackled the critical problem of explaining how the human species had acquired mental powers that were, even by Darwin's admission, lifted far above the level enjoyed by our closest animal relatives. There are many passages in the book which reveal Darwin's commitment to a progressionist view of evolution, even though he knew that the human race could not be seen as the intended goal of the evolutionary process. But he was sufficiently imbued with the logic of his own branching model of development that he recognized the inability of simple progressionism to explain the origin of the human mind. Many evolutionists believed that mental progress was simply inevitable; given enough time life was bound to ascend to the human level. From this perspective it would be counterproductive to suggest that there was something unique about our ancestors which allowed them to develop a greatly enhanced intelligence. Darwin knew that such a way of thinking was a trap which evaded the real problem of human origins. If mental progress is inevitable, why should the apes not have kept up with our ancestors and arrived at the same level as ourselves? In a branching model of evolution it is essential to specify why the two branches diverging from a common starting point have moved in different directions. Paradoxically Darwin's theory required him to specify unique conditions affecting our ancestors, because only in this way could he explain the difference in the mental powers acquired by humans and apes.

In effect Darwin had to provide what modern evolutionists call an 'adaptive scenario' to explain why our ancestors had evolved characters that separated them from the apes. His solution was to switch attention from our mental powers to another unique human character: our upright posture and bipedal means of locomotion. Darwin realized that this represented an adaptation to life in a different kind of environment to that favoured by the apes. The apes have remained apes because they have retained their ancestral lifestyle in the trees, and their forelimbs have thus continued to be adapted for grasping branches. Our own ancestors moved out of the trees and stood upright as a means of getting about on the open plains. This in turn freed their hands for exploring the environment and for using sticks and stones as primitive tools.[11] Darwin thus implies that our intelligence is a by-product of a unique shift in lifestyle by our ancestors. In their new way of life, natural selection favoured those individuals who walked upright and in turn promoted the increase of intelligence within a population that now had better opportunity to exploit that faculty.

Darwin's suggestion as to the reason why our ancestors stood upright is not accepted by modern anthropologists, but it was a pioneering effort to grapple with the problem of explaining why the line of human evolution became separated from that of the apes. Later discoveries in the fossil record have at least confirmed that Darwin was right to suspect that the earliest humans stood upright before the brain began to increase in size. Few of his contemporaries were prepared to admit that the brain did not lead the way in human evolution. For the most part they preferred to believe that standing upright was a consequence of increased intelligence, not its cause. As in so many other areas, the majority of nineteenth-century evolutionists were so imbued with faith in the inevitability of progress that they were unable to recognize the significance of the problem that Darwin was attempting to solve.

Darwin was convinced that our moral sense was a product of the interaction between the social instincts and the developing intelligence. He did not believe that increased socialization was responsible for the growth of intelligence because he knew that the apes and many other animals already live in family groups. He argued that, in animals with such a lifestyle, evolution would naturally produce instincts that encouraged cooperative behaviour and even some level of self-sacrifice. In part this would result from what is now known as 'group selection': because of the advantages conferred by cooperation, those groups with more strongly developed social instincts would displace others in which those instincts were less powerful. Darwin also invoked an element of Lamarckism, arguing that the inherited effects of social habits would result in the creation of instincts. He noted that in many 'primitive' tribes the willingness to cooperate with others is confined to the tribal group – outsiders do not count as part of the moral universe. This was consistent with the view that the social instincts were built up for the benefit of the group. Only as human intelligence has expanded has it become possible for us to rationalize our instinctive feelings by postulating absolute moral standards. As the size of our societies has increased we have inevitably been led to generalize the moral imperatives forced upon us by evolution, creating religious and moral theories designed to convince us that respect for others is an absolute good.

These were powerful claims, and we can see why religious thinkers (and even some evolutionists) were reluctant to carry the theory to such lengths. But in other respects Darwin was a child of his own time. Whatever his initial feelings about slavery he was now convinced that the coloured races had lagged behind the whites in the ascent from the apes. He accepted measurements given by several contemporary authorities according to which the average brain capacity of the white race was larger than that of any other. He assumed that a larger brain meant a greater level of intelligence, thus placing Europeans at the head of a hierarchy of racial types. Like most of his contemporaries, Darwin became convinced that the Europeans were conquering the world not just because they had superior technology but because they were brighter than the other races. He commented at length on the various factors which seemed to drive the 'lower' races into a decline towards extinction when confronted by white colonists. In this respect, if no other, Darwin was a 'social Darwinist'. He still disapproved of overtly harsh treatment of blacks but he accepted the inevitability of white supremacy.

On the other hand Darwin was convinced that all human races share a common ancestry and belong to the same species. There were some attempts to argue that the various races had originated from different ape ancestors and were thus distinct species. Even Wallace suggested that the races had branched apart from one another in the remote past and had advanced independently towards the modern human form. But

Darwin knew that the races could all interbreed successfully with one another, and he accepted that this implied a level of divergence that was insufficient to generate distinct species. He was also suspicious of the claim that several related lines could independently advance in the same direction. Such a degree of parallel evolution would imply a directing agent incompatible with the theory of natural selection.

Darwin was prepared to accept that some races have advanced more rapidly in the area of mental progress, perhaps because they have been exposed to a more challenging environment. But he remained puzzled by many of the physical specializations that characterize the modern races. Perhaps the negroes' black coloration was somehow correlated with their bodies' resistance to tropical diseases, but Darwin found it difficult to see how other racial characteristics could be explained in terms of adaptation. In the end he decided that the only explanation was sexual selection, and for this reason the second half of the *Descent of Man* is devoted to this topic. If particular characters had become associated with sexual attractiveness in certain populations, those characters would be enhanced because individuals possessing them to a high degree would reproduce more often than average. A classic example was that of the Hottentot women in whom 'the posterior part of the body projects in a wonderful manner' a peculiarity which is 'greatly admired by the men'. According to the traveller Richard Burton, the Somal men were also supposed 'to choose their wives by ranging them in a line, and by picking her out who projects farthest *a tergo*. Nothing can be more hateful to the negro than the opposite form.'[12] By extending this line of argument Darwin could account for the non-adaptive characters that distinguish all the various races by attributing them to different ideas of beauty.

Social Evolutionism

The *Descent of Man* tied the human race firmly into Darwin's vision of a world governed by natural evolutionary forces. But by the time the book appeared the idea of evolution was already being applied to the

development of culture and society by anthropologists and archaeologists. As late as the 1850s it was still widely believed that the human
race had appeared only a few thousand years ago. Reports that primitive
stone tools had been found in ancient deposits along with the bones of
extinct animals were dismissed as fraudulent. With such a limited
timespan for human history it was easy to believe the biblical story that
Adam had been taught the arts of civilization by his Creator. Any form
of cultural evolutionism in which society progressed from an originally
primitive state was unthinkable. Darwin's own speculations about
primitive human origins, and the more public expression of the same
position in Chambers' *Vestiges*, were very much at variance with contemporary beliefs. But in the late 1850s the situation began to change
dramatically; archaeologists at last began to accept that the human race
had a considerable antiquity and that the oldest remains indicated a
very primitive level of technology. Cultural and social progress now
became the key to prehistory, and an evolutionary account of how the
most primitive humans had appeared began to seem more plausible.[13]

Historians of archaeology used to think that the sudden acceptance
of a vast antiquity for the human race was a by-product of the debate
over the *Origin of Species*. But recent studies suggest that the emergence
of cultural evolutionism was an independent development that would
have taken place whether or not Darwin had published his theory.
Lyell's *Antiquity of Man* of 1863 summed up the new evidence, although
we have seen that Lyell himself was hostile to an evolutionary account
of human origins and was still suspicious of the basic idea of biological
evolution. Evidence for the existence of a stone age as the earliest stage
in prehistory had been building up for some time, and it was almost
inevitable that the growing popularity of a progressionist account of
society would force a reassessment of the evidence. The Darwinian
revolution in biology and the emergence of cultural evolutionism were
parallel developments that coincided in time – although the two ideas
were soon linked by the suggestion that the most primitive stone-age
humans had evolved from ape-like ancestors. In the *Descent of Man*
Darwin was able to draw upon a decade of rapid expansion in archaeology and anthropology to substantiate his view of primitive human
origins.

Perhaps the clearest evidence that cultural evolutionism was not merely a spin-off from its biological equivalent can be seen in the very non-Darwinian character of the developmental model used by the archaeologists and anthropologists.[14] Where Darwin saw biological evolution as a branching tree, the cultural evolutionists constructed a ladder of developmental stages which, they assumed, all races of mankind could ascend. Archaeologists such as John Lubbock divided the stone age into a progressive sequence of stages defined by improvements in toolmaking. They also looked to modern 'savages' as illustrations of how our remote ancestors would have lived. Cultures with primitive technologies were automatically assumed to have preserved the social structure that the Europeans' ancestors had passed through in the stone age. In effect, modern primitives were relics of the stone age, cultural fossils preserved in areas where the environment did not provide sufficient stimulus for them to advance further up the scale. Social evolution was thus portrayed as a linear progressive sequence with modern industrial civilization as the goal towards which all races were, at least potentially, aspiring. Anthropologists such as Edward B. Tylor used this developmental model in their study of non-European societies, trying to determine the point that each society had reached in the progressive scale.

Darwin himself drew on this progressionist model in the *Descent of Man* in order to create the impression that mankind must have advanced from very primitive origins. He knew that the fossil record gave little evidence for the emergence of mankind from the apes – hence the popularity of the term 'missing link' to denote the intermediate stages of development. The evidence for the primitive state of early technology was used as a substitute for the missing fossils in order to build up the case for the undeveloped mental state of our earliest toolmaking ancestors. Lubbock was a close neighbour of the Darwin family at Down, and Darwin read his *Prehistoric Times* (1865) with enthusiasm.[15] To the extent that the *Descent of Man* made use of the archaeologists' progressionist model of prehistory, it deflected attention away from the crucial question of why our earliest ancestors had diverged from those of the apes. The developmental scheme postulated in cultural evolution simply assumed that progress was aimed at a single goal (the creation of modern civilization) and used environmental factors merely to explain why some races had advanced further up the scale than others.

Anyone locked into this way of thinking would find it difficult to appreciate the significance of Darwin's brief suggestion of an adaptive scenario to explain why human evolution was turned into its unique path of mental development.

It is significant that Darwin's heir-apparent in the field of mental evolution, George John Romanes, adopted an explicitly developmental approach that paid little or no attention to the question of why the human race had ascended so much further than the apes. Romanes was a young physiologist who worked on the nervous system of invertebrates.[16] He first met Darwin in the summer of 1874 and the two men soon struck up a close relationship. Romanes conceived the project of extending the argument of the *Descent of Man* to give a more complete account of mental evolution and Darwin supported him in this. Romanes' *Animal Intelligence* of 1881 gave more evidence for the well-developed mental powers of the higher animals. Following Darwin's death, his *Mental Evolution in Animals* and *Mental Evolution in Man* attempted to outline the sequence in which the mental powers had been developed. *Mental Evolution in Animals* contained as an appendix an essay on instinct based on material that Darwin had originally intended to include in the *Origin of Species*.

In fact Romanes' accounts of mental evolution owed more to the philosophy of Herbert Spencer than to Darwin's biological theory. His approach was to trace out a logically plausible sequence by which the mental functions of animals with the simplest nervous system could be developed through to the human level of intelligence. Romanes paid little attention to the question of why evolutionary pressures should have generated each successive advance – like Spencer he assumed that progress was more or less inevitable. He certainly made no attempt to follow up Darwin's suggestion that a change of lifestyle and habitat was crucial for turning human evolution into a path that would separate it from the apes. Although recognizing that natural selection could act on instincts, Romanes preferred Spencer's Lamarckian approach in which instincts were produced when learned habits became so deeply ingrained that they became hereditary. His whole approach implied that the purpose of evolution was to promote the development of intelligence towards the human level. Given this essentially teleological framework, it is small wonder that Romanes retained an interest in formal religion throughout his career and, after passing through an agnostic phase, returned eventually to an acceptance of Christianity.

Post-Darwinian studies of mental and social evolution thus provided parallel lines of support for an essentially progressionist view of the origins of humankind. Darwin's brief recognition that his theory of branching evolution made it necessary to specify unique circumstances leading to the establishment of a distinctively human line of evolution was obscured by his own and his followers' preference for a system in which progress towards higher levels of mentality was taken for granted. But what did this progressionist synthesis imply for the social policies of the time? The critics of Darwinism – then and now – imply that the theory opens the floodgates that will allow all traditional moral values to be swept away by a ruthless worship of brute force. The claim that the late nineteenth century was dominated by an amoral 'social Darwinism' has been made by many historians and is still supported by modern critics such as the creationists. Yet our study of Darwin and his followers hardly supports such an interpretation. The evolutionists certainly accepted that mankind had evolved from the brutes and they realized that our distant ancestors would have lived without the moral values we take for granted today. But they had no intention of portraying nature as a purposeless system that provided no foundation for moral values. Evolutionism was designed to provide a new foundation for morality, not to sweep morality away. Nature itself was designed to promote the growth of those values that we hold dear, and thus we can feel pride in our position as the advance guard in the inevitable advance towards higher things.

The harsher side of this progressionism can be seen in its treatment of the 'lower' races of mankind. Anthropologists such as Lubbock were quite convinced that modern 'savages' retained a primitive level of culture because their mentality remained at a lower level than the more advanced whites. They were living fossils both culturally *and* mentally, preserving an earlier stage in biological evolution through into the modern world. Savages were, in effect, the missing links in the ascent of man from the apes – they were visible in the modern world even if their fossil equivalents had not yet been discovered. This position had already become apparent in Spencer's evolutionism even before Darwin published the *Origin of Species*. Spencer realized that biological evolutionism made it possible to argue that the development of the mind and the development of civilization went hand in hand. The more intelligent races developed a higher culture and this culture in turn stimulated further mental progress. On this model the uncivilized races must be presumed to have a lower intelligence. The Victorians were confident that their industrial progress indicated a higher level of intelligence for the white race and they were anxious to find excuses for their conquest of other peoples. Even without the concept of the 'struggle for existence', evolutionism allowed non-industrialized societies to be identified as the products of a primitive mentality. The 'lower' races were thus evolutionary failures: whatever the cause, they had lagged behind in the march of progress and would never be able to catch up. Their displacement by the more advanced types was only a matter of time.

We have seen that Darwin shared this attitude towards the lower races, although he disapproved of any deliberate cruelty towards the inferior branches of the human stock. In the end he thought that the replacement of lower by higher types was inevitable, except in those areas where whites could not penetrate because the conditions were unsuitable. Towards the end of his life he wrote to a correspondent about the struggle for existence among races and concluded: 'Looking to the world at no very distant date, what an endless number of the lower races will have been eliminated by the higher civilized races throughout the world.'[17] It is easy to see how others might use the theory of the 'survival of the fittest' to justify a more ruthless attitude towards the lower races. Even if natural selection were only a negative factor designed to eliminate those branches of evolution which failed to progress, it seemed to legitimize the policy of displacing other races from territory which – by the standards of industrial society – they could not exploit properly. The paradox of progressionism was that a belief in the purposeful character of evolution could be combined with a ruthless attitude towards nature's failures. It was precisely because evolution was designed to produce higher levels of mentality that it became necessary to eliminate those who did not keep up. There were, in fact, many non-Darwinian evolutionists who were quite willing to accept the negative role of selection in the elimination of undeveloped races.

But what about competition among the inhabitants of industrialized Europe? The classic image of 'social Darwinism' is that of a ruthless individualism in which the logic of *laissez-faire* capitalism is pushed to its utmost extreme. The struggle for existence is glorified as the means by which the lazy and stupid are eliminated so that the most able individuals can lead the way towards progress. The Darwinian system does clearly contain elements that could be used to support such an amoral view of the world. Success would be the only criterion of 'good' in a world governed purely by natural selection, and success at any price would be the social message such a theory would convey. According to some historians Darwinism did indeed usher in a materialistic ethic in which all traditional values were swept away by a worship of brute force. Herbert Spencer is often portrayed as the leading 'social Darwinist' – a man carried away with enthusiasm both for evolutionism and for the free-enterprise system. Spencer is seen as the leading architect of a new ideology for capitalism which allowed successful businessmen to invoke Darwin's theory as a means of justifying their ruthlessness by claiming that the 'survival of the fittest' led to progress.[18]

But did anyone at the time actually invoke Darwinism to justify struggle as a means of eliminating unfit members of their own society? Modern studies suggest that it is by no means easy to find clear-cut examples of 'social Darwinism' in the late nineteenth century.[19] Certainly no one used the actual phrase 'social Darwinism' at the time – it was introduced as a term of abuse in the early twentieth century. More seriously, the supposed link between the logic of Darwin's theory and contemporary justifications of capitalism is difficult to establish. It is certainly true that Darwin drew some of his inspiration from the individualism that characterized the society in which he lived, and was well read in the classics of *laissez-faire* economics, including Adam Smith and, of course, Malthus. But Darwin's use of the individualist model was highly original and did not reflect the typical justification for that view of society adopted by the contemporary middle classes. The whole point of the progressionist ideology of social evolutionism was that it did *not* destroy all moral values, leaving room only for a free-for-all struggle. Progress could only be defined by assuming that there was a morally significant goal for evolution to attain, and the actual process of evolution itself had to be seen as an expression of moral purpose. The ethical standards of individualism may have been harsh, but its supporters nevertheless believed that they were upholding a new foundation for certain traditional values.

The Darwinians themselves did not see their theory as a licence to justify unrestrained competition within society. In the *Descent of Man* Darwin noted that the civilized nations have circumvented the ability of natural selection to eliminate the unfit by instituting poor laws, medical care and other ways of helping the unfortunate. He admitted that this was potentially harmful to the race since – by analogy – no breeder of domestic animals would allow his worst animals to reproduce. Yet Darwin pointed out that we are charitable to the poor because we have social instincts built into us by natural selection, and he went on to argue that even today there is a natural tendency for the immoral to die without reproducing.

> In regard to the moral qualities, some elimination of the worst dispositions is always in progress even in the most civilized nations. Malefactors are executed, or imprisoned for long periods, so that they cannot freely transmit their bad qualities. Melancholic and insane persons are confined, or commit suicide. Violent and quarrelsome men often come to a bloody end. The restless who will not follow any steady occupation – and this relic of barbarism is a great check to civilisation – emigrate to newly-settled countries, where they prove useful pioneers.[20]

Thus, far from promoting sheer ruthlessness, natural selection is still helping to develop the cooperative instincts which lie at the basis of morality. In a postscript to a letter to Lyell in 1860 Darwin wrote: 'I have received, in a Manchester newspaper, rather a good squib, showing that I have proved "might is right," and therefore that Napoleon is right, and every cheating tradesman is right.'[21] The 'squib' certainly shows how the potential for 'social Darwinism' could be read into Darwin's theory, but his reaction is hardly that of someone who really wished to advocate that 'might is right' – or even of someone deeply concerned that he would be interpreted in this way by the majority of his readers.

Of Darwin's supporters, Wallace was openly opposed to the capitalist system and advocated socialism on the grounds that the accumulation of wealth distorted the natural choice of marriage partners. Huxley eventually abandoned his support for social progressionism and became one of the leading opponents of Spencer's philosophy. His lecture 'Evolution and Ethics' of 1893 argued against what eventually became known as social Darwinism on the very Darwinian grounds that natural evolution is not necessarily progressive and hence cannot serve as a reliable guide in human affairs.[22] The suggestion that Darwinism naturally arose from a commitment to ruthless individualism is thus rendered implausible by the views of the Darwinians themselves.

Even Spencer's argument for individualism as the key to social evolution can be seen to have non-Darwinian foundations. Since it was Spencer, not Darwin, who applied the term 'survival of the fittest' to denote the action of natural selection, it is easy to imagine that Spencer himself was a Darwinian in science.[23] His advocacy of an extreme *laissez-faire* position in which the state takes no responsibility for alleviating the miseries of the unfit can appear to be based on the view that progress depends on the elimination of these unfortunates. Yet Spencer's main argument for individualism was based on a very different mechanism that has far more in common with the Lamarckism which he saw as the main driving force of animal evolution. The purpose of struggle was not to eliminate the unfit but to force them to become fitter. The miseries attendant upon failure were the best possible way of educating the lazy so that they would be more industrious and more enterprising in future. Eliminating congenital stupidity was only a secondary factor – the vast majority of people had the ability to function adequately in the world if only they put their minds to it, and the advantage of unrestrained individualism was that it forced everyone to maximize their efforts and exploit their initiative to the full. Evolution for the species came about through the accumulation of many generations of individual self-improvement, which Spencer assumed would be transmitted from parent to offspring through education. Eventually the habit of self-reliance would become so deeply ingrained that it would be instinctive. Thus Spencer's 'social Darwinism' turns out to be really a form of 'social Lamarckism'.

Darwin and Spencer used the concept of the struggle for existence in quite different ways: where Darwin postulated selection acting upon a random variation of hereditary characters, Spencer thought most individuals could transcend the limits of inheritance if stimulated by the threat of suffering. Yet both were convinced that evolution has a moral purpose. Darwin saw selection as a force that enhanced the social instincts lying at the heart of morality, while Spencer thought that nature rewarded hard work, thrift and initiative as a means of building those characteristics into the very nature of humanity. The suffering of the unfit was an unfortunate by-product, a necessary harshness designed to eliminate the unfit characters in future generations. In effect Spencer's theory took the traditional virtues of the Protestant work-ethic and built them into nature as the driving force of progressive evolution. Far from being rejected as a materialist because of his scepticism about formal religion, Spencer was welcomed by some Protestant liberals precisely because his system provided a new foundation for traditional middle class values.[24] The claim that 'Darwinism' – whether in its original or in its Spencerian form – precipitated Victorian culture into an age of moral nihilism fails to take account of the faith in the inevitability of progress that served as the foundation for social evolutionism.

To become a strict Darwinian in human affairs one would have to insist that all aspects of personality are rigidly determined by heredity, and that progress can only come about through the constant elimination of those individuals unfortunate enough to be born with inferior characteristics. Such a hereditarian approach to social issues was indeed suggested by Darwin's cousin, Francis Galton, in his *Hereditary Genius* of 1869. Galton distrusted Darwin's theory of pangenesis because he preferred to believe that heredity is a much more rigid force determining how parental characters are transmitted to the offspring. He argued that heredity was of paramount importance in human affairs and attempted to substantiate this position by showing how superior intelligence tended to run in families. Galton's suggestion for improving the human race was based on the application of what was, in effect, artificial selection: he proposed that superior people (the professional classes) should be encouraged to have more children, while the inferior masses huddled in the slums should be somehow prevented from breeding. Eventually he coined the term 'eugenics' to denote such a policy and proposed measures to identify and institutionalize those feeble-minded individuals who were the worst threat to the genetic future of the race.[25]

Darwin discussed Galton's ideas in the *Descent of Man* and seems to have accepted that his proposals offered the only realistic prospect of deliberately improving the human race. Significantly, however, he saw that there were major obstacles standing in the way of identifying those who most deserved to be eliminated from the breeding stock.

> Though I see so much difficulty, the object seems a grand one; and you have pointed out the sole feasible, yet I fear utopian, plan of procedure in improving the human race. I should be inclined to trust more (and this is part of your plan) to disseminating and insisting on the importance of the all-important principle of inheritance.[26]

This letter to Galton suggests that Darwin did recognize the hereditarian principles at the heart of his theory, but he was less than optimistic about the prospect of the state taking control of human breeding. As a child of the age of individualism he saw persuasion rather than compulsion as the only path to progress.

Significantly, Galton was largely ignored during the 1870s and 1880s; the Victorians were simply not ready for a view of society in which everyone's abilities were so rigidly determined. They preferred Spencer's appeal to nature's hardships as a means of stimulating the unfit to improve themselves. Only at the very end of the century did Galton find that his views were beginning to gain a hearing, and the eugenics movement was to become a significant part of the political scene in the early decades of the twentieth century. Its most extreme manifestation was as part of the Nazis' attempts to purify the Aryan race. The logic of eugenics was, no doubt, inspired originally by Darwin's theory, but its belated acceptance as a serious programme suggests that we can hardly blame Darwin for generating this particular form of social Darwinism. The scientific and social developments that at last allowed Galton to influence public affairs span several highly eventful decades. Spencer's *laissez-faire* social evolutionism lost credibility in the face of mounting calls for state control in order to defend the empire's position in the world, while in science the advent of Mendelian genetics did much to focus attention on the role of heredity. If eugenics is truly a form of social Darwinism, the fact that it did not succeed in Darwin's own lifetime confirms that the selection theory was not merely an expression of the mid-Victorian capitalist ethos.

11

Darwin and the Modern World

B Y THE TIME of Darwin's death in 1882 the theory of
evolution had become almost universally accepted by
scientists. Through its absorption into the progressionist ideology the
general idea of evolution became one of the dominant themes of
Victorian thought. Darwin had become a cultural symbol, the figurehead for one of the major intellectual developments witnessed by his
contemporaries. Yet his own views on how evolution worked remained
highly controversial. Far from triumphing over the alternatives, the
theory of natural selection seemed unable to deal with the objections
that had been raised against it and an increasing number of biologists
were invoking alternatives such as Lamarckism. The late nineteenth
century saw an 'eclipse of Darwinism' in which many scientists set
themselves up in conscious opposition to the selection theory. Darwin
had brought about a revolution despite the fact that his own views were
significantly at variance with those of the majority of his 'followers'. The
Victorians had difficulties with natural selection because they could not
accept the idea of undirected evolution. Darwin's theory had been published at the right time to tip the balance in favour of exploring the
general concept of evolution, but most of his followers decided that
there must be something more purposeful than natural selection in

control. For every evolutionist who followed Darwin in the study of adaptation and biogeography there were dozens who preferred to concentrate on the grand sweep of progress up to mankind revealed by the fossil record.

It is an irony of history that the theory of natural selection should be revived decades after Darwin's death to become the foundation for modern evolutionism. The early exponents of Mendelian genetics themselves rejected the selection theory but they had undermined the more developmental alternatives proposed in the late nineteenth century and soon came to realize that selection was the only directing force they could still invoke. By the 1940s the revival of Darwinism was well under way and now there was less temptation for biologists to see the theory as a contribution to progressionism. The twentieth century has adopted a more pessimistic view of human affairs and has thus been able to take on board the full logic of the open-ended model of evolution that Darwin suggested. To blame our loss of faith on Darwin seems a little harsh considering that he, like most of his contemporaries, struggled to see a cosmic purpose in evolution that would preserve at least some traditional values. As an event in cultural history the Darwinian revolution belongs firmly to the nineteenth century. The problems that beset our own times are a product of the circumstances that led Western civilization to reject the Victorian faith in progress that Darwinism did much to create.

Darwin thus remains a difficult figure for us to evaluate. The aspects of his theory that modern biologists see as most innovative are precisely those which had to be suppressed in his own time – and which Darwin himself found it impossible to confront in human affairs. The Darwinian revolution falls into two distinct stages; the conversion of the Victorian world to evolutionism, and the revival of the selection theory in modern times. What, then, should we mean by 'Darwinism'? Is it the theory of open-ended divergent evolution brought about by natural selection that modern biologists find so exciting, or is it the compromise that Darwin and his contemporaries favoured in which evolution is the mechanism of inevitable progress? There is no simple answer to this question. Darwin drew upon and appealed to the attitudes of his own time, yet he also exploited a new line of evidence that forced him to see evolution in a light that only a few of his fellow biologists could appreciate. Darwin is a major figure in the history of science because in at least some respects he transcended the values of his own time to create an idea that had the potential for development in later decades. We should never forget the extent to which he shared the common attitudes of Victorian Britain, but if we pretend that he did nothing more than project those attitudes onto nature we miss the spark of creativity which gave his theory permanent value.

The Death of Darwin . . .

The 1870s saw a gradual improvement in Darwin's health, allowing him to continue work on the botanical projects described above in chapter 8. For all the profound interest generated by the idea of progressive evolution Darwin himself chose to study the small-scale effects that were most directly illuminated by his theory of local adaptation due to natural selection. Others might prefer to explore the origin of new classes and the ascent of life on earth but Darwin recognized that such grand problems were almost beyond the scope of scientific evolutionism, given the limited availability of fossils. By studying the plants he could grow in his own garden Darwin knew that he could throw light on the process of adaptation that lay at the heart of his theory. Natural selection was a complex process that resulted from the interaction of a population with its environment; it turned nature into a law-bound system, but at the same time made that system so complex that detailed predictions about the course of evolution became almost impossible. If Darwin retained a faith in the ultimately progressive character of evolution, he was wise enough to realize that it was only in the long term that such a trend would become apparent. His decision to work on the problem of adaptation reflects the kind of theory he had proposed and illustrates the differences between that theory and the confident progressionism preferred by most Victorian evolutionists.

The book on earthworms appeared in 1881 and was Darwin's last major publication. He began a series of experiments on the chemicals responsible for producing insect galls in plants, but this project was never completed. There was renewed controversy over the use of vivisection in physiological experiments and Darwin wrote a letter to *The Times* defending the scientists concerned against the charge of causing unnecessary suffering.[1] He was instrumental in providing financial security for Wallace who was now supporting himself by writing. Having heard that Wallace was in difficulties Darwin prepared a Memorial to Gladstone asking for a civil list pension. He obtained the signatures of Huxley, Hooker and a number of other prominent scientists on the Memorial, and Gladstone responded by awarding Wallace a pension of £200 per annum. Aware that his time was running out Darwin also decided that some of his own considerable fortune (boosted by a series of prudent investments) should be devoted to a general project of value to naturalists. Through Hooker he contributed money to be used in the compilation of an index of the names of all the genera and species of plants, eventually to be known as the *Index Kewensis*.[2]

Not surprisingly he was now the recipient of many honours. He had been elected to the membership of many foreign scientific societies and was awarded a number of honorary degrees including an LLD from Cambridge in 1877.[3] Gladstone offered to make him a trustee of the British Museum but he had to decline on the grounds of ill health. It is perhaps worth noting that the once popular story that Karl Marx offered to dedicate a volume of *Capital* to Darwin is based on a misinterpretation of the relevant correspondence.[4]

Darwin did not grow more religious in his declining years and there is no evidence to support the story circulated by some fundamentalists that he underwent a deathbed conversion. He did not believe in revelation and was doubtful about an afterlife. He remained convinced that it was impossible to believe in a God who personally superintended every event in the material universe. On the question of whether there might be some more general element of design in the world he remained confused. The Duke of Argyll (a prominent exponent of theistic evolutionism) recalled that he broached this subject with Darwin during the last years of his life:

> In the course of that conversation I said to Mr. Darwin, with reference to some of his own remarkable works on the 'Fertilisation of Orchids,'

and upon 'The Earthworms,' and various other observations he made of the wonderful contrivances for certain purposes in nature – I said it was impossible to look at these without seeing that they were the effect and the expression of mind. I shall never forget Mr. Darwin's answer. He looked at me very hard and said, 'Well, that often comes over me with overwhelming force; but at other times,' and he shook his head vaguely, adding, 'it seems to go away.'[5]

In a letter of 1879 Darwin wrote: 'In my most extreme fluctuations I have never been an atheist in the sense of denying the existence of God. I think generally (and more and more as I grow older), but not always, that an Agnostic would be the more correct description of my state of mind.'[6]

Although he had lost some of the symptoms of his earlier illness Darwin began to lose vigour in his last years. He became tired very easily and in his last letter to Wallace he wrote:

> We have just returned home after spending five weeks on Ullswater; the scenery is quite charming, but I cannot walk, and everything tires me, even seeing scenery ... What I shall do with my few remaining years of life I can hardly tell. I have everything to make me happy and contented, but life has become very wearisome to me.[7]

He began to suffer from heart pains early in 1882. During the night of April 18th he suffered a severe attack and was only brought back to consciousness with difficulty. Francis Darwin records that he said 'I am not the least afraid to die'.[8] He was faint and nauseous through the following morning and died at about four o'clock.

The family had assumed that he would rest at Down but John Lubbock almost immediately began a campaign to have him buried in Westminster Abbey.[9] A letter signed by Lubbock and nineteen other MPs was sent to the Dean of Westminster, and Lubbock also wrote to the family asking that they agree to the suggestion. Darwin was buried in the Abbey on 26 April, the pallbearers including Lubbock, Huxley, Wallace, Hooker and the Duke of Argyll. The grave is in the north aisle of the nave, close to those of Newton, Faraday and Lyell. Burial in the Abbey was the one national honour that Darwin received and it

PLATE 15 *Charles Darwin* by John Collier, 1881 (reproduced by kind permission of the Darwin Museum, Down House, courtesy of The Royal College of Surgeons of England)

can be interpreted as an effort to symbolize the cultural transformation brought about by his theory. Far from destroying all traditional values evolutionism had merely transferred the responsibility for safeguarding those values from the Church to the professional scientists. That, at least, was the message that Lubbock and Huxley would have liked to convey.

... and the Rebirth of Darwinism

At the time of Darwin's death his theory of natural selection was entering a period of temporary decline. The decades around 1900 saw a growth in the popularity of non-Darwinian mechanisms of evolution and an increased willingness among scientists to proclaim themselves openly hostile to 'Darwinism'. There were of course celebrations to mark the fiftieth anniversary of the *Origin of Species* in 1909 – Wallace and Hooker were still alive to provide a semblance of continuity with the original debate. But Darwin was hailed only as the founder of evolutionism, most biologists having lost confidence in his mechanism of natural selection. Even the emergence of Mendelian genetics in the years after 1900 seemed at first to offer only another alternative to selectionism. But Mendelism turned out to be even more fundamentally opposed to Lamarckism and the other theories of directed variation. Eventually the geneticists began to realize that selection by the environment was the only factor that could control the flow of genes within a population. Darwinism was reborn in a new guise, largely purged of the progressionist associations that had been an integral part of Victorian evolutionism.[10]

It would be wrong to pretend that there was no continuity linking Darwin's work to the emergence of modern Darwinism. Some nineteenth-century Darwinists had appreciated the special nature of the theory and had followed Darwin's lead in the study of those areas of biology that could be most readily illuminated by it. Biogeography was a major focus of attention in the careers of evolutionists such as Wallace and Hooker. Others studied the process of local adaptation which formed the cutting edge of natural selection's operations. Wallace's *Darwinism* of 1889 provided a clear and comprehensive survey of the theory and of the relevant areas of biology. Except in the case of the origin of the human mind, Wallace was an extreme selectionist; unlike Darwin, he would have nothing to do with any other mechanism of evolution. This position soon became known as 'neo-Darwinism' to distinguish it from the more flexible form of the theory which Darwin himself had advocated and which had gained support precisely because it allowed selection to be relegated to the status of a secondary mechanism.

Even here, though, there were technical problems that stood in the way of the theory's development. In his early career the experience of the Galapagos had taught Darwin that geographical isolation was an important factor in speciation (the splitting up of a single parent species into a number of 'daughter' species). But by the time he wrote the *Origin of Species* he had become convinced that geographical barriers were not essential; ecological specialization could split a population in two even when there was no geographical barrier preventing inter-breeding across the intermediate territory.[11] Later Darwinists found it difficult to explain how such a separation could occur, and one reason for the popularity of theories based on saltations or sudden large-scale variations was that these could account for the creation of a reproductively isolated population. Only when renewed studies of biogeography began to confirm that a physical barrier to interbreeding *was* essential did it become possible for a later generation of Darwinists to solve the problem of speciation.

Such technical problems dovetailed with the general preference for some form of predetermined variation that would give evolution a 'direction' other than that imposed by local adaptation. The 1880s and 1890s saw many biologists opting openly for Lamarckism, orthogenesis or saltationism, especially in areas such as palaeontology where the lessons of Darwinism were particularly difficult to apply. We have already noted that even Francis Darwin became reconciled to the Lamarckian ideas of Samuel Butler, despite the quarrel with his father. Julian Huxley, grandson of T. H. and an architect of the modern synthesis of Darwinism and genetics, looked back on this period as the 'Eclipse of Darwinism', recognizing that the theory had been at its lowest ebb at the turn of the century.[12] The crisis of confidence was caused by a number of factors: the Darwinists' failure to resolve outstanding problems; the ease with which non-Darwinian mechanisms could be applied in other areas such as palaeontology; and the increasing dogmatism of the neo-Darwinists who were now refusing to compromise with alternative theories. Those biologists who preferred the alternatives were now forced to declare themselves opponents of Darwinism.

One topic that proved especially controversial was heredity. Darwin had been willing to allow a role for Lamarckism because he had found it impossible to break with the old viewpoint in which the question of how characters are transmitted from parent to offspring is subordinated to the more general problem of how the offspring's body grows from the material supplied by the parents (the egg and sperm). It was almost impossible for a nineteenth-century biologist to conceive of a study that would concentrate solely on the transmission of characters, bypassing the problem of how those characters were developed in the growing organism. In such a climate of opinion it was very difficult to think of characters as being discrete units transmitted unchanged from one generation to the next. Lamarckism, orthogenesis and saltationism flourished because even the Darwinists admitted that *development* was an important aspect of how variations are produced and transmitted. Changes in the growth pattern were the raw material of evolution and it was easy to imagine that the processes controlling growth were somehow responsible for directing the course of variation – and hence the course of evolution. Since *development* is by definition goal directed, it was possible to imagine that evolution was also directed towards some preordained goal.

Darwin's theory of pangenesis was a classic product of this developmental model of variation and heredity. Instead of challenging the model head-on, Darwin worked within it and was thus prevented from seeing what a modern Darwinist would regard as the most obvious implication of his theory. He realized that natural selection threw the emphasis onto those characters that were preserved intact by heredity, but he was incapable of visualizing those characters as units that could be transmitted independently of the process by which they are produced in the growing organism. It has been claimed that if Darwin could have read Gregor Mendel's reports of his experiments on heredity in peas published in 1865, the problems of the selection theory could have been overcome immediately. But such a claim rests firmly on hindsight: we know that Mendel's experiments became the foundation for modern genetics, and we assume that Darwin and his followers would have been able to appreciate the significance that we now read into those results. History tells us that no one at the time was able to appreciate the implications of what Mendel was doing – indeed it now seems almost certain that even Mendel himself did not think of his experiments as the basis for a new theory of heredity.[13] Had he known of Mendel's results, Darwin would probably have dismissed them as an interesting anomaly. He would hardly have abandoned pangenesis because a single experiment could not have outweighed a lifetime's training in a different thought pattern.

Darwin escaped the more blatant applications of the developmental viewpoint because his experience with the breeders taught him that variation was random, that is, undirected. Whatever the cause of variation it did not impose a preordained direction on the course that evolution could take. He was thus able to develop his theory of open-ended divergent evolution which made predicting the future course of change impossible because it subordinated everything to the demands of local adaptation and hence to the hazards of migration and environmental change. But pangenesis allowed for the inheritance of acquired characters as well as for random variation, and this was why Darwin failed to confront the possibility that characters could be transmitted as units. The majority of his contemporaries preferred to believe that the process of individual growth did somehow direct variation along preordained channels thus reducing selection to a secondary mechanism and giving evolution a direction imposed from within the organism, independently of environmental pressures. Selection would remain just one mechanism among many as long as the prevailing view of heredity and variation allowed for the possibility of such a directing force. The degree of Darwin's success – and failure – in influencing his contemporaries was determined by his retention of the developmental model of heredity. Since he did not challenge that model, his ideas could easily be treated as a contribution to the progressionist view of evolution. This allowed the selection theory to play a vital role as catalyst in the transition to evolutionism but prevented anyone from seeing the truly revolutionary implications of the theory. Even Darwin only dimly grasped the fact that his approach had the potential to undermine the whole logic of progressionism.

The late nineteenth century saw a focusing of attention on the problem of heredity that paved the way for the emergence of Mendelism. Two biologists played a key role although they approached the problem from very different directions. In Germany, August Weismann criticized pangenesis from the viewpoint of the newly developed science of cytology, the microscopic study of cell structure. Weismann now insisted that the cell nucleus was responsible for transmitting the information from the parents that determined the structure of the growing organism. He insisted that the nucleus of the reproductive cells could not be influenced by changes going on within the parent's body. The substance of heredity, which he called the 'germ plasm', was transmitted unchanged from one generation to the next. This brought the hereditarian implications of Darwinism into sharper focus. As Weismann correctly realized, his vision of the germ plasm's role made Lamarckism impossible; variation must be the product of changes taking place purely within the germ plasm, and natural selection alone can direct the way in which new characters can build up within the population. Weismann became the archetypal neo-Darwinist, proclaiming the 'all-sufficiency of natural selection' and forcing all biologists who remained faithful to Lamarckism into outright opposition.[14]

The other major contributor was Darwin's cousin, Francis Galton. We have already seen in chapter 10 how Galton became convinced that the human personality is rigidly determined by heredity. To prove his point he set out to study the role of heredity, but instead of looking at the cellular processes responsible for fertilization he chose to investigate the effects of heredity and variation in large populations. He developed statistical techniques to allow the study of character distribution in populations over many generations. Although Galton himself favoured saltations to originate new species, his approach focused attention on the population and helped to clarify an important point of relevance to Darwin's theory. Like most of his contemporaries Darwin had tended to visualize heredity and variation as two antagonistic processes: heredity was responsible for accurately copying parental characters in the offspring, while variation interfered with this process to produce differences. Galton's technique allowed him to see that in fact heredity and variation can be treated as two different aspects of the same process, provided we analyse them in terms of the population rather than the individual. The population exhibits variability because a wide range of fixed hereditary factors are being circulated within it, combined and recombined in different ways due to individual matings.

Galton's disciple, Karl Pearson, extended his statistical work and decided that the new approach to heredity was compatible with Darwin's theory of natural selection. Selection would act on the range of hereditable variation in the population, gradually shifting the range in the direction of increased fitness. His co-worker, W. F. R. Weldon, made field studies designed to show the action of natural selection in producing small but definite changes in the range of variability when a population was exposed to a changed environment. At the height of the 'eclipse of Darwinism', Pearson's biometrical school kept alive a faith in the possibility of natural selection as an answer to the problem of evolution.

Neither Weismann nor Galton saw their hereditary factors as distinct units that might be traced through several generations by breeding experiments. Pearson was actively hostile to the idea that discrete characters could be traced because he saw the variation of a population as a continuous range, with most individuals clustered around the mid-point of the range. But in the closing years of the century several factors coincided to make possible the emergence of a new science based on the identification of unit characters, a science that would at first be known as Mendelism but which would finally take on its modern name, genetics. There was immense controversy over whether or not the structure of the growing organism was predetermined by the germ plasm it inherited, but gradually the balance was tipping in favour of those who claimed that it was. At the same time the study of variation was leading a few biologists towards the construction of breeding experiments that could trace characters through a number of generations. In 1900, two botanists independently 'rediscovered' the results that Gregor Mendel had published in 1865. Carl Correns and Hugo De Vries showed that certain plant characters can be treated as distinct units that are visible over many generations, even though in some cases the 'recessive' version of a character is completely masked by the 'dominant' form.

A leading exponent of the new science, and the man who provided the first English translation of Mendel's paper, was William Bateson. At the start of his career Bateson had been trained as an evolutionary morphologist and had attempted to reconstruct the origin of the vertebrates by studying the embryology of some of the lowest modern forms. He eventually gave this up in disgust, realizing that the fossil record was unlikely to yield evidence that would allow his conclusions to be tested. Darwin himself would have discouraged this kind of evolutionary investigation for precisely that reason, and the fact that Bateson turned against 'Darwinism' illustrates just how far the definition of that term had been modified from what Darwin had intended. In the hope of turning evolution into an experimental science Bateson began to study variation and he soon became interested in characters that seemed most likely to have been produced instantaneously rather than by the accumulation of minute differences. He attacked the selection theory vigorously, claiming that the environment had no power to interfere with the production of new characters by saltations. In the hope of identifying the distinct characters produced in this way he began to set up breeding experiments and was thus well prepared to acknowledge the significance of Mendel's laws when they were announced by Correns and De Vries.

Far from being hailed as the salvation of Darwinism, then, the new science of genetics was founded by biologists who were convinced that evolution took place by the sudden appearance of discrete new characters. De Vries went on to create his 'mutation theory' based on what he thought were experimentally observable saltations in the evening primrose. Bateson had no time for Darwinism, and Pearson, whose biometrical approach was firmly committed to the natural selection of continuously varying characters, rejected the discrete units of the Mendelians as irrelevant. And yet the advent of Mendelism had transformed the whole situation in a way that was to have immense implications for the fortunes of Darwin's theory. Attention was now firmly fixed on the ability of heredity to transmit characters in an absolutely fixed manner from one generation to the next. The geneticists were no longer interested in the question of how the characters were manifested in individual development and they were thus not inclined to see growth as a force directing evolution. Lamarckism and orthogenesis were inconceivable within their view of heredity: evolution consisted of the production of new genetic characters by mutation, and mutation was a process that could not be controlled by events taking place during the growth of the embryo. With the exception of saltationism, the alternatives to Darwinism had been routed. Evolution was the process by which new genetic characters were introduced into the population. The only questions remaining were: how large are the new characters, and does selection by the environment have any control over how new mutations spread into the population?

Over the next few decades the geneticists gradually realized that large mutations are invariably harmful. If evolution is to make use of new characters produced in this way, it is likely to be only the smaller changes that will have a chance to be perpetuated. Biologists such as R. A. Fisher, J. B. S. Haldane and Sewall Wright developed the science of population genetics to study the range of genetic factors responsible for maintaining the variability of a large population. Fisher was one of Pearson's students but he soon realized that the statistical approach of biometry would have to take account of the genetical processes responsible for the transmission of unit characters. He showed that the continuous range of variation studied by the Darwinians was actually sustained by the circulation of a number of genetic factors, each of which had some effect on a particular character. In most cases we cannot see the individual Mendelian units because their effect is blended together. Fisher also showed how natural selection could alter the range of variation by enhancing the reproduction of those genes responsible for producing a favourable character within a particular environment. By the time Fisher's *Genetical Theory of Natural Selection* of 1930 was published the gulf between genetics and Darwinism was beginning to disappear. Far from destroying Darwinism, genetics had saved it by eliminating the alternatives and by showing that classic objections such as Fleeming Jenkin's 'swamping' argument were based on a false view of heredity.[15]

Fisher and Haldane paid no attention to the geographical dimension of evolution. They postulated a form of natural selection acting uniformly on a large population, with no allowance made for what might happen if a small group of organisms was transported to a new location. Sewall Wright's model of selection, however, was based on the idea that a species might be divided into small partially isolated reproductive groups. At the same time field naturalists such as Ernst Mayr and Julian Huxley were becoming more interested in the problems of biogeography and were beginning to suspect that Darwin's original Galapagos-inspired view of speciation was correct. Species were formed when samples of an existing population were transported to new locations where they formed reproductively isolated groups. It now began to look increasingly likely that natural selection was the process responsible for adapting each group to its new environment, gradually building up so much change that the isolated population would not be able to interbreed with the parent form even if the geographical barrier were removed. Theodosius Dobzhansky's *Genetics and the Origin of Species* of 1937 was important for its efforts to translate the abstruse mathematics of population genetics into practical conclusions that could be tested by the field naturalists. Mayr's *Systematics and the Origin of Species* and Huxley's *Evolution: the Modern Synthesis*, both of 1942, are now counted as classics in the foundation of the modern synthetic form of Darwinism. George Gaylord Simpson's *Tempo and Mode in Evolution* of 1944 argued that the new model could be applied to palaeontology, displacing the Lamarckian and orthogenetic theories that had predominated in that field since the end of the previous century.

Although there have been many challenges to its authority in recent decades, the synthesis of Darwinism and genetics remains the dominant paradigm in evolutionary thought. Biologists now take it for granted that evolution is a process driven by the demands of local adaptation, that it is an open-ended divergent process with no 'main line' of development and that there is no purely biological force driving organisms towards higher levels of development. The last vestiges of nineteenth-century progressionism, still visible in the work of Julian Huxley, have been eliminated. It was no accident that the 1940s also saw the acceptance of a very 'Darwinian' view of human origins, breaking at last with the assumption that evolution must somehow be driven towards the human level of intelligence. Fossil Australopithecines from southern Africa confirmed Darwin's original prediction that our ancestors stood upright *before* the great expansion of the human brain began. The separation of the human family was the product of an adaptive modification, not the inevitable outcome of a trend towards bigger brains. Far from being a predictable goal of evolution the human race now appears to be the unlikely outcome of a chapter of lucky accidents. Of all the nineteenth-century evolutionists Darwin was the only one who even partially anticipated such a radical conclusion.

The anti-progressionism of modern Darwinism seems quite in tune with the less confident attitude of a culture that has been shaken by two world wars and which has begun to suspect that its own technological triumphs are no guarantee of spiritual progress. But what role did Darwin's theory play in the loss of faith that is so characteristic of modern times? Critics such as the creationists blame evolutionism for most of the world's ills. They argue that if we reduce mankind to the level of the brutes and base our idea of evolution on the 'survival of the fittest', there must inevitably be a collapse of traditional moral values. But did Darwin's theory really play such a decisive role in undermining faith in the purposeful character of the universe which produced us? We have seen that, in his own time, evolutionism succeeded only because it pandered to the prevailing faith that nature *was* progressive and hence purposeful in its actions. Even Darwin found it impossible to escape the prevailing faith in progress although he realized that his theory created major problems for any concept of a necessarily progressive force. The supporters of Darwinism saw it as a reformulation of moral values, not as a force that would destroy those values altogether. Only the opponents of evolutionism alleged that it would destroy religious faith.

Where Darwinism did pave the way towards modern values was in the support it appeared to provide for a more ruthless attitude towards failure. Such an attitude would have emerged in Victorian times without Darwin's help of course. This is confirmed by the fact that Spencer's evolutionary view of society was based more on Lamarckian than on Darwinian principles. But the theory of natural selection certainly provided ammunition for those who wished to argue that it was 'only natural' that the weak should be eliminated. Darwin and Spencer certainly shared this view in the case of those human races which they thought had been left behind by the march of progress. By the end of the century the idea of a 'struggle for existence' among nations was being used to justify the imperialism that was leading Europe ever closer to war. Yet we should be careful to recognize the other factors that were involved in the rise of militarism. As liberals, Darwin and Spencer were opposed to the idea that the military should play a large role in a nation's affairs. Spencer was appalled at the direction taken by European culture in the late Victorian age. The period leading up to the outbreak of World War I covered the 'eclipse of Darwinism', when the selection theory was at its lowest ebb in science. Many of the concepts exploited by the ideology of imperialism came from sources that had always been opposed to the liberalism within which Darwin's theory was conceived.[16] Darwin cannot altogether be blamed for the use of catch-phrases such as 'the survival of the fittest' by militarists who had no appreciation of the real logic of his theory. At best Darwinism was just one among many of the factors that drew European civilization ever closer to a war that would undermine faith in the inevitability of progress once and for all.

We still hear the term 'social Darwinism' used even today. But more often than not it is applied to ideologies that have only indirect connections with those of Darwin's time. Anyone who claims that human nature is in part determined by genetic inheritance is likely to be called a social Darwinist, even though the hereditarianism of genetics has only indirect links to the original form of Darwinism. There have certainly been plenty of non-Darwinians who have claimed that the various human races have different levels of mental ability. The Darwinian influence is clearest perhaps in the new science of sociobiology, which seeks to explain animal instincts in terms of natural selection. Sociobiologists such as Edward O. Wilson have certainly claimed that human behaviour is also influenced by genetically imprinted instincts.[17] Yet the great triumph of sociobiology has been to explain apparently altruistic instincts in terms of Darwinian natural selection. Through the concept of the 'selfish gene' it is possible to show that natural selection can build up instincts that will lead the individual to sacrifice its own interests for those of others. The claim that such instincts affect human behaviour may justifiably be termed a form of social Darwinism, but it is a version that bears little resemblance to the old image of a society based on ruthless competition. Those who oppose genetic determinism in its varying forms seem to find the term 'social Darwinism' a useful label for all that they find abhorrent in modern biology. But they would do well to recognize that there are many different sources of the philosophy they dislike so much, not all of which can be traced back to Darwin's time.

The evolutionists of the late-Victorian era had their own reasons for hailing Darwin as a hero of science. Lubbock, Huxley and the others who ensured that he was buried in Westminster Abbey saw this honour as appropriate for the man who had played a decisive role in allowing science to tackle an area of study that had hitherto remained off-limits. Darwinism was the symbol of the progressionism at the heart of a new ideology that made science the chief source of authority in Western culture. It is doubtful whether many of the people who attended Darwin's funeral thought that his theory of natural selection would remain at the centre of evolutionary biology. Modern biologists agree that Darwin deserves his place in the Abbey but they do so for very different reasons. Darwin is remembered, where Spencer is largely forgotten, because his efforts to grapple with the problems of bio-geography led him to formulate a theory that was able to transcend the progressionism of his time. Modern science honours Darwin both as the focus for the debate which established evolutionism as an integral part of biology and as the founder of the mechanism which is still regarded as the most plausible explanation of how evolution works. It may be difficult for us to believe that the reasons why his theory succeeded in its own time were not the same as those which now appeal to us. To visualize Darwin in his own context we must remember that his contemporaries were unable to appreciate precisely those aspects of his thinking which seem most important today. He was both a product of his own time and a thinker who created an idea capable of being exploited by later scientists with very different values. Any attempt to understand Darwin the man must first take into account the multiple roles that his name has played within the symbolism of both nineteenth- and twentieth-century thought.

Notes

Chapter 1

1 Moore, 'Charles Darwin Lies in Westminster Abbey', 1982. Moore's *Post-Darwinian Controversies* exposes the weakness of the metaphor of a war between science and religion, as expressed, for instance, in White, *A History of the Warfare of Science with Theology in Christendom*, 1896.

2 See Lovtrup, *Darwinism: the Refutation of a Myth*, 1987.

3 On American creationism see Nelkin, *The Creation Controversy*, 1983. For an example of the non-fundamentalist opposition to the materialism of the selection theory see Koestler, *The Ghost in the Machine*, 1967.

4 See Caplan (ed.), *The Sociobiology Debate*, 1978. Young's *Darwin's Metaphor*, 1985, offers a good illustration of a socially conscious historian's efforts to re-evaluate the impact of Darwinism.

5 De Beer, *Charles Darwin*, 1963.

6 E.g. Clark, *The Survival of Charles Darwin*, 1985.

7 Janet Browne of the Darwin Letters Project at Cambridge University Library is working on a detailed biography.

8 A good example of an account stressing Darwin's creativity is Gruber, *Darwin on Man*, 1974.

9 E.g. Lovtrup, *Darwinism: the Refutation of a Myth*, 1987.

10 Mayr, *The Growth of Biological Thought*, 1982.

11 Barzun, *Darwin, Marx, Wagner*, 1958; Himmelfarb, *Darwin and the Darwinian Revolution*, 1959.

12 Bowler, *The Non-Darwinian Revolution*, 1988.

Chapter 2

1 For details of pre-Darwinian evolution theories see Bowler, *Evolution: The History of an Idea*, 1989, chs 3 and 5, and Greene, *The Death of Adam*, 1959.

2 See Desmond, *Archetypes and Ancestors*, 1982; 'Robert E. Grant', 1984; 'Artisan Resistance and Evolution in Britain, 1819–1848', 1987; and *The Politics of Evolution*, 1989.

3 See Hodge, 'Lamarck's Science of Living Bodies', 1971; Burkhardt, *The Spirit of System*, 1977; Jordanova, *Lamarck*, 1984; and Corsi, *The Age of Lamarck*, 1988.

4 Darwin, *Autobiography* (ed. Barlow), 1958, p. 49.

5 On Grant's influence on the young Darwin see Sloan, 'Darwin's Invertebrate Program', 1985.

6 Grant, 'Observations on the Nature and Importance of Geology', 1826. On Grant's influence in Edinburgh and London see Desmond, *The Politics of Evolution*.

7 Secord, 'Behind the Veil', 1989.

8 Millhauser, *Just before Darwin*, 1959; see also Gillispie, *Genesis and Geology*, 1951, ch. 6.

9 Hodge, 'The Universal Gestation of Nature', 1972.

10 On the complex role of phrenology, see Cooter, *The Cultural Meaning of Popular Science*, 1985.

11 Chambers, *Vestiges of the Natural History of Creation*, 1844; see the chapters on 'Early History of Mankind' and 'Mental Constitution of Animals'.

12 Sedgwick, 'Vestiges of the Natural History of Creation', 1845.

13 Owen, *On the Nature of Limbs*, 1849. On Owen and evolution see E. Richards, 'A Question of Property Rights', 1987.

14 Corsi, *Science and Religion: Baden Powell and the Anglican Debate*, 1988.

15 See Peel, *Herbert Spencer: The Evolution of a Sociologist*, 1971. On Spencer as a 'social Darwinist' see Hofstadter, *Social Darwinism in American Thought*, 1959.

16 Spencer's 'The Development Hypothesis' is reprinted in his *Essays*, 1983, vol. I, pp. 381–7. On the Lamarckian character of his social and biological evolutionism see Bowler, *The Non-Darwinian Revolution*, 1988, chs 2 and 7.

17 On Huxley see Di Gregorio, *T. H. Huxley's Place in Natural Science*, 1984, and Desmond, *Archetypes and Ancestors*.

18 Huxley, 'Vestiges of the Natural History of Creation', 1853.

19 Huxley, 'On the Reception of the Origin of Species', 1887, pp. 196–7.

20 On nineteenth-century palaeontology see Rudwick, *The Meaning of Fossils*, 1972, and Bowler, *Fossils and Progress*, 1976.

21 On Buckland see Rupke, *The Great Chain of History*, 1983, and more generally on the reinterpretation of catastrophist geology see Hooykaas, *Natural Law and Divine Miracle*, 1959.

22 See Rehbock, *The Philosophical Naturalists*, 1983.

23 On the impact of Geoffroy's thinking see Appel, *The Cuvier–Geoffroy Debate*, 1987.

24 Owen, 'Report on British Fossil Reptiles, Part 2', 1841; on Owen's position see Desmond, 'Designing the Dinosaurs', 1979; *Archetypes and Ancestors*; and 'Richard Owen's Reaction to Transmutation in the 1830s', 1985.

25 Sedgwick, 'Vestiges of the Natural History of Creation'; Miller, *The Old Red Sandstone*, 1841, and *Footprints of the Creator*, 1850. See Bowler, *Fossils and Progress*, ch. 4.

26 On Darwin's response to *Vestiges* see Egerton, 'Refutation and Conjecture', 1970.

27 Owen, *On the Nature of Limbs*, 1849, p. 89.

28 See Ospovat, 'The Influence of Karl Ernst von Baer's Embryology', 1976, and Bowler, *Fossils and Progress*, ch. 4.

Chapter 3

1 Darwin, *The Life and Letters of Charles Darwin*, 1887, vol. I, ch. 2. Henceforth the *Life and Letters* will be abbreviated as LLD and *More Letters of Charles Darwin*, 1903, as MLD.

2 *The Autobiography of Charles Darwin*, 1958 (henceforth abbreviated as *Autobiography*).

3 Darwin, *A Calendar of the Correspondence of Charles Darwin*, 1985, and *The Correspondence of Charles Darwin*, 1985–8 (henceforth abbreviated as *Calendar* and *Correspondence*).

4 Darwin, *Charles Darwin's Notebooks*, 1987 (henceforth abbreviated as *Notebooks*).

5 Hodge, 'Darwin as a Lifelong Generation Theorist', 1985.

6 See for instance King-Hele, *Erasmus Darwin*, 1963, and McNeil, *Under the Banner of Science*, 1987. On the Darwin family ethos see also H. E. Gruber, *Darwin on Man*, 1974, ch. 3.

7 Hodge, 'Darwin as a Lifelong Generation Theorist'.

8 LLD, vol. I, p. 11.

9 See the letter from Catherine and Caroline Darwin, *Correspondence*, vol. I, p. 41.

10 LLD, vol. I, p. 32, and *Autobiography*, p. 27.

11 Schweber, 'The Wider British Context in Darwin's Theorizing', 1985.

12 Sloan, 'Darwin's Invertebrate Program', 1985. The zoology notes from Darwin's Edinburgh notebooks are reprinted in *Notebooks*, pp. 477–86.

13 LLD, vol. I, p. 41, and *Autobiography*, p. 52.

14 LLD, vol. I, p. 45, and *Autobiography*, p. 57.

15 Letter from Fanny Owen, January 1828, *Correspondence*, vol. I, p. 48.

16 Darwin to Caroline Darwin, 28 April 1831, *Correspondence*, vol. I, p. 122. On Sedgwick's involvement in the election see Sedgwick, *Life and Letters*, 1890, vol. I, pp. 373–6.

17 LLD, vol. I, p. 47, and *Autobiography*, p. 59.

18 On Sedgwick's career see his *Life and Letters*; Secord, *Controversy in Victorian Geology*, 1986; and Rudwick, 'A Year in the Life of Adam Sedgwick and Company', 1988. For a negative assessment of catastrophist geology see Gillispie, *Genesis and Geology*, 1951; for more recent (and more sympathetic) accounts, see Hooykaas, *Natural Law and Divine Miracle*, 1959, and Rudwick, 'Uniformity and Progression', 1971.

19 LLD, vol. I, p. 57, and *Autobiography*, p. 69.

20 Sedgwick to Darwin, 4 September 1831, *Correspondence*, vol. I, pp. 137–8.

21 Darwin to Henslow, 11 July 1831, *Correspondence*, vol. I, p. 125.

22 Letters relating to the *Beagle* opportunity appear in the *Correspondence*, vol. I, pp. 127–43.

23 *Correspondence*, vol. I, pp. 133–4.

24 Frederick Watkins to Darwin, 18 September 1831, *Correspondence*, vol. I, p. 159.

25 This episode is described in the autobiography, LLD, vol. I, pp. 59–60, and *Autobiography*, p. 72.

Chapter 4

1 LLD, vol. I, p. 61, and *Autobiography*, p. 76.

2 E.g. Moorehead, *Darwin and the Beagle*, 1969; Keynes, *The Beagle Record*, 1979; and Ralling, *The Voyage of Charles Darwin*, 1978. For a general history of biogeography see Browne, *The Secular Ark*, 1983.

3 Parts of Darwin's notebooks and some letters from the voyage are transcribed in Darwin, *Charles Darwin and the Voyage of the Beagle*, 1945. His *Beagle* diary is available in a 1933 edition edited by Nora Barlow and in a new edition, by R. D. Keynes, 1988. The complete letters from the voyage appear in volume I of the *Correspondence*. Extracts from the diaries and letters along with original pictures by the *Beagle* artists appear in Keynes, *The Beagle Record*, 1979.

4 *Correspondence*, vol. I, p. 237.

5 Ibid., p. 436.

6 Sloan, 'Darwin's Invertebrate Program', 1985.

7 Sulloway, 'Darwin's Early Intellectual Development: an Overview of the *Beagle* Voyage', 1985.

8 On Darwin's collections, see Porter, 'The *Beagle* Collector and his Collections', 1985.

9 Darwin, *Journal of Researches*, 1891, p. 9.

10 To Caroline Darwin, 24 October – 24 November 1832; *Correspondence*, vol. I, pp. 277–8.

11 *Journal of Researches*, p. 14.

12 *Autobiography*, pp. 73–5.

13 Ibid., p. 85.

14 To Caroline Darwin, 20 September 1833; *Correspondence*, vol. I, pp. 330–1.

15 *Journal of Researches*, p. 59.

16 See Rachootin, 'Owen and Darwin Reading a Fossil', 1985.

17 *Journal of Researches*, p. 125.

18 Ibid., p. 150.

19 25 October 1833; *Correspondence*, vol. I, pp. 344–5.

20 For the Henslow letters see ibid., pp. 292–4 and 327–8, and for Darwin's replies, pp. 369 and 397–402.

21 Ibid., p. 398.
22 Ibid., p. 401.
23 Darwin to Henslow, ibid., p. 443.
24 Darwin to W. D. Fox, ibid., p. 460.
25 *Journal of Researches*, p. 276.
26 Sulloway, 'Darwin and his Finches: the Evolution of a Legend', 1982.
27 *Journal of Researches*, p. 287.
28 *Correspondence*, vol. I, p. 499.
29 23 April 1836; ibid., p. 495.

Chapter 5

1 See for instance the letter from Caroline Darwin to Elizabeth Wedgwood, *Correspondence*, vol. I, pp. 504–5.
2 Rudwick, 'Charles Darwin in London', 1982.
3 Darwin to Henslow, 1 October 1836, *Correspondence*, vol. I, p. 512.
4 Ibid.
5 Letters to and from both the government and the publisher may be found in *Correspondence*, vol. II, pp. 23–46.
6 See Rudwick, 'Charles Darwin in London'. See also Herbert, 'Darwin the Young Geologist', 1985.
7 E.g. *Correspondence*, vol. II, pp. 13 and 52.
8 Darwin's 'Observations on the Parallel Roads of Glen Roy' appears in his *Collected Papers*, 1977, vol. I, pp. 87–137. See Rudwick, 'Darwin and Glen Roy', 1974.
9 *Autobiography*, p. 84; LLD, vol. I, p. 68.
10 Darwin to Henslow, 14 October 1837, *Correspondence*, vol. II, pp. 50–2. The original letter to William Whewell turning down the offer is ibid., pp. 9–10.
11 Darwin's evaluation of the scientific and literary figures he met can be found in the *Autobiography*, pp. 98–113; LLD, vol. I, pp. 68–78.
12 *Correspondence*, vol. II, p. 284.
13 See Ruse, 'Darwin's Debt to Philosophy', 1975.
14 The note and the paper are reproduced in *Correspondence*, vol. II, pp. 443–5.
15 Ibid., pp. 171–2. On Darwin's home life see Moore, 'Darwin of Down', 1985.
16 See the letters to Henslow, 20 September 1837 and 14 October 1837, *Correspondence*, vol. II, pp. 47–8 and 50–2.
17 For a survey of the debate on Darwin's illness see Colp, *To be an Invalid*, 1977.
18 Oldroyd, 'How did Darwin Arrive at his Theory?', 1984.
19 See Rudwick, 'Charles Darwin in London', and Sulloway, 'Darwin's Early Intellectual Development', 1985.
20 *Correspondence*, vol. II, p. 238.
21 Ibid., pp. 357–79, and replies from Waterhouse, pp. 379–82.
22 These are printed in ibid., pp. 446–9.

23 On contemporaries' ideas see chapter 2 above. For further comments on Darwin's originality see Bowler, *The Non-Darwinian Revolution*, 1988, ch. 2. For a survey of current thinking on the development of Darwin's ideas see Hodge and Kohn, 'The Immediate Origins of Natural Selection', 1985.

24 See Hodge, 'Darwin and the Laws of the Animate Part of the Terrestrial System', 1982.

25 Sulloway, 'Darwin's Early Intellectual Development'.

26 Hodge, 'Darwin as a Lifelong Generation Theorist', 1985. Sloan also notes Darwin's involvement in a debate on the properties of living matter; see his 'Darwin, Vital Matter, and the Transformism of Species', 1986.

27 See Secord, 'Darwin and the Breeders', 1985.

28 See Ospovat, *The Development of Darwin's Theory*, 1981.

29 On the relationship between Darwin and Malthus see for instance Young, 'Malthus and the Evolutionists', 1969, reprinted in Young, *Darwin's Metaphor*, 1985; and Bowler, 'Malthus, Darwin, and the Concept of Struggle', 1976.

30 The link with Paley's approach was stressed in Cannon's 'The Bases of Darwin's Achievement', 1961.

31 Kohn 'Darwin's Ambiguity: the Secularization of Biological Meaning', 1989. On Darwin's theism see Brown, 'The Evolution of Darwin's Theism', 1986.

32 For an interpretation of Darwin as a progressionist, see for instance R. J. Richards, *Darwin and the Emergence of Evolutionary Theories of Mind and Behaviour*, 1987.

Chapter 6

1 To Caroline Darwin, 24 July 1842, *Correspondence*, vol. II, pp. 324–5.

2 To W. D. Fox, 29 April 1851, LLD, vol. I, p. 380. See also ibid., pp. 132–4.

3 See Francis Darwin's 'Reminiscences of My Father's Everyday Life', LLD, vol. I, ch. 3.

4 To Emma Darwin, 20 or 27 October 1844, *Correspondence*, vol. III, p. 68.

5 Moore, 'Darwin of Down', 1985.

6 *Autobiography*, p. 115; LLD, vol. I, p. 79.

7 Secord, 'Darwin and the Breeders', 1985.

8 To Hooker, 11 January 1844, *Correspondence*, vol. III, p. 2. On Hooker's life and career see his *Life and Letters*, 1918 (ed. L. Huxley).

9 LLD, vol. I, p. 142.

10 The 'Sketch' and 'Essay' are printed in Darwin and Wallace, *Evolution by Natural Selection*, 1958.

11 Ospovat, *The Development of Darwin's Theory*, 1981.

12 *Correspondence*, vol. III, pp. 43–4.

13 See Egerton, 'Refutation and Conjecture', 1970.

14 MLD, vol. I, p. 38.

15 LLD, vol. I, pp. 144–50.

16 To Hooker, 10 September 1845, *Correspondence*, vol. III, p. 233.

17 See Ghiselin, *The Triumph of the Darwinian Method*, 1969, ch. 5.

18 Kohn, 'Darwin's Principle of Divergence as Internal Dialogue', 1985. See also Browne, 'Darwin's Botanical Arithmetic', 1980.

19 See Sulloway, 'Geographic Isolation in Darwin's Thinking', 1979.

20 *Autobiography*, p. 93.

21 Darwin and Wallace, *Evolution by Natural Selection*, pp. 114–16.

22 Young, Darwin's Metaphor: Does Nature Select?', 1971. Also reprinted in Young, *Darwin's Metaphor*, 1985.

23 See MLD, vol. I, p. 114 n. 2.

24 To Hooker, 30 December 1858, MLD, vol. I, p. 115.

25 See Ospovat, *The Development of Darwin's Theory*, and Brown, 'The Evolution of Darwin's Theism', 1986.

26 See Stauffer's introduction to *Charles Darwin's Natural Selection*, 1975.

Chapter 7

1 On Wallace's life see his autobiography, *My Life*, 1905, and the *Letters and Reminiscences*, 1916 (ed. A. Marchant). Biographies include George, *Biologist–Philosopher*, 1964; Fichman, *Alfred Russel Wallace*, 1981; McKinney, *Wallace and Natural Selection*, 1972; and Williams-Ellis, *Darwin's Moon*, 1958. See also Beddall, *Wallace and Bates in the Tropics*, 1969.

2 Wallace's 1855 and 1858 papers are reprinted in his *Contributions to the Theory of Natural Selection*, 1870, and *Natural Selection and Tropical Nature*, 1895.

3 The joint Darwin–Wallace papers are reprinted in Darwin and Wallace, *Evolution by Natural Selection*, 1958.

4 See Brackman, *A Delicate Arrangement*, 1980, and Brooks, *Just Before the Origin*, 1983.

5 See Bowler, 'Alfred Russel Wallace's Concepts of Variation', 1976, and Nicholson, 'The Role of Population Dynamics in Natural Selection', 1960. For a commentary on the relationship between the two scientists see Kottler, 'Charles Darwin and Alfred Russel Wallace', 1985.

6 See the *Origin of Species* variorum text edited by Morse Peckham. Note also the concordance to the first edition of the *Origin of Species*.

Chapter 8

1 Darwin to Wallace, 13 November 1859, LLD, vol. II, p. 221.

2 To Hooker, 25 November 1859, ibid., pp. 225–6.

3 Ibid., vol. III, p. 31.

4 See MLD, vol. II, p. 9. The Second Wrangler is the candidate placed second in the results of the Mathematics Tripos at Cambridge.

5 *Autobiography*, p. 138; LLD, vol. I, pp. 100–1.

6 LLD, vol. II, p. 371. On the scientific debate see Bowler, *Evolution: The History of an Idea*, 1984, ch. 7; Hull (ed.), *Darwin and his Critics*, 1973; Ruse, *The Darwinian Revolution*, 1979, ch. 8; and Vorzimmer, *Charles Darwin: The Years of Controversy*, 1970.

7 To Huxley, 27 November 1859, LLD, vol. II. p. 282.

8 Huxley to Darwin, 23 November 1859, LLD, vol. II, pp. 231–2.

9 Ibid.

10 See the list, ibid., vol. II, p. 293.

11 To Lyell, 1 June 1860, ibid., vol. II, p. 315.

12 See ibid., vol. III, pp. 27–8.

13 Huxley to Darwin, 12 September 1868, *Life and Letters of Huxley*, 1908, vol. I, p. 428. 'Darwinismus' is the German for 'Darwinism'.

14 On the late developments in the theory of pangenesis see Geison, 'Darwin and Heredity', 1969, and Olby, 'Charles Darwin's Manuscript of Pangenesis', 1963. On the origins of Darwin's views on the subject see Hodge, 'Darwin as a Lifelong Generation Theorist'.

15 Hodge, 'Darwin as a Lifelong Generation Theorist', 1985.

16 Eiseley, *Darwin's Century*, 1958, ch. 8. For a critique of this interpretation see Bowler, 'Darwin's Concepts of Variation', 1974.

17 Jenkin, 'The Origin of Species', 1867; also reprinted in Hull (ed.), *Darwin and his Critics*.

18 Wallace to Darwin, 2 July 1866, MLD, vol. I, pp. 269–70.

19 LLD, vol. I, pp. 147–9. On the background to the botanical work see the letters in LLD, vol. III, chs 7–12, and MLD, vol. II, chs 10 and 11.

20 LLD, vol. III, p. 255.

21 Ibid., vol. III, p. 290.

22 To Hooker, 21 November 1860, ibid., vol. III, p. 319.

23 Ibid., vol. III, p. 290.

24 23 September 1860, ibid., vol. II, p. 343.

25 Hull, 'Darwinism as a Historical Entity', 1985.

26 See the letters in MLD, vol. I, ch. 6.

27 See Kottler, 'Charles Darwin and Alfred Russel Wallace', 1985.

28 Huxley, 'On the Reception of the "Origin of Species"', 1887, p. 197.

29 See Desmond, *Archetypes and Ancestors*, 1982, and Di Gregorio, *T. H. Huxley's Place in Natural Science*, 1984.

30 Huxley, 'Evolution in Biology' (1878); reprinted in *Collected Essays*, 1893–4, vol. II, pp. 187–226 at p. 223.

31 Huxley, *American Addresses*, 1888, pp. 85–90; also reprinted in *Collected Essays*, vol. IV, see p. 132. On the fossil evidence for evolution see Bowler, *Fossils and Progress*, 1976; Desmond, *Archetypes and Ancestors*; and Rudwick, *The Meaning of Fossils*, 1972.

32 For a critique of the metaphor of a 'war' between science and religion, see Moore, *The Post-Darwinian Controversies*, 1979.

33 For a reassessment of the Oxford debate, see Jensen, 'Return to the Huxley–Wilberforce Debate', 1988.

34 See Desmond, *Archetypes and Ancestors*. Huxley had at first been suspicious of the idea of biological progress, and these suspicions revived at the end of his career, but in the 1860s he threw in his lot with the liberal progressionists.

35 See MacLeod, 'The X-Club', 1970.

36 See *Life and Letters of Hooker*, 1918, vol. II, ch. 35.

37 *Life and Letters of Huxley*, vol. II, p. 67.

38 See Wallace, *Alfred Russel Wallace*, 1916, pp. 257–8.

Chapter 9

1 We have already encountered this interpretation in chapter 8 above; see Eiseley, *Darwin's Century*, 1958, and more recently Mayr, *The Growth of Biological Thought*, 1982.

2 For hostile accounts of Darwinism's influence, see for instance Himmelfarb, *Darwin and the Darwinian Revolution*, 1959, and Barzun, *Darwin, Marx, Wagner*, 1958. For a detailed survey of the scientific criticisms see Vorzimmer, *Charles Darwin: The Years of Controversy*, 1970. On the religious debates see Moore, *The Post-Darwinian Controversies*, 1979.

3 This incident was reported by the entomologist Roland Trimen; see Poulton, *Charles Darwin and the Origin of Species*, 1909, pp. 213–14.

4 See Jensen, 'Return to the Huxley–Wilberforce Debate', 1988.

5 For a collection of reviews see Hull (ed.), *Darwin and his Critics*, 1973. On the debate in the periodical press see Ellegard, *Darwin and the General Reader*, 1957.

6 *Autobiography*, pp. 93–4 (the remarks on religion were omitted from the version of the autobiography printed in LLD).

7 See Brown, 'The Evolution of Darwin's Theism', 1986.

8 Gray, *Darwiniana*, 1876, p. 148. See Dupree, *Asa Gray*, 1959.

9 Darwin, *Variation of Animals and Plants under Domestication* (2nd edn, 1882), vol. I, p. 428.

10 See Sedgwick's letter to Darwin, 24 December 1859, LLD, vol. II, pp. 247–50. Sedgwick also wrote a hostile review in the *Spectator*; see ibid., vol. II, pp. 297–8.

11 Ellegard's *Darwin and the General Reader* shows that many periodical articles in the late 1860s accepted evolution but opted for a more purposeful mechanism than natural selection.

12 Herschel, *Physical Geography*, 1861, p. 12. On the 'higgledy-piggledy' comment see Darwin to Lyell, 12 December 1859, LLD, vol. II, p. 241. The contrast between Herschel's search for a timeless order in nature and Darwin's open-ended view of historical development is stressed in Schweber, 'John Herschel and Charles Darwin', 1989.

13 On Owen and evolutionism see E. Richards, 'A Question of Property Rights', 1987.

14 Owen, 'On the Origin of Species', 1860.

15 To Lyell, 5 December 1859, LLD, vol. II, p. 240.

16 9 April 1860, MLD, vol. I, pp. 145–7.

17 Owen, *Anatomy of the Vertebrates*, 1868, vol. III, p. 808.

18 For details of Kelvin's arguments see Burchfield, *Lord Kelvin and the Age of the Earth*, 1975, and on Darwin's response, Burchfield, 'Darwin and the Dilemma of Geological Time', 1974.

19 On Mivart see J. W. Gruber, *A Conscience in Conflict*, 1960; Vorzimmer, *Charles Darwin: The Years of Controversy*; and Desmond, *Archetypes and Ancestors*, 1982, ch. 4.

20 To Wallace, 30 January 1871, LLD, vol. III, p. 135.

21 To Hooker, 16 September 1871, MLD, vol. I, p. 333.

22 On the emergence of orthogenesis and other anti-Darwinian theories of evolution in the late nineteenth century see Bowler, *The Eclipse of Darwinism*, 1983.

23 Shaw attacked Darwinism in the Preface to *Back to Methuselah*; for Koestler's critique see *The Ghost in the Machine*, 1967.

24 Spencer's 'The Development Hypothesis' is reprinted in his *Essays*, 1883, vol. I, pp. 381–7.

25 To Hooker, 10 December 1866, LLD, vol. III, pp. 55–6. On Spencer see Peel, *Herbert Spencer*, 1971, and on the link between Darwinism and Spencer's philosophy see Greene, *Science, Ideology and World View*, 1981, especially ch. 6.

26 See Spencer, *Principles of Biology*, 1864, vol. I, p. 444. For Darwin's comments on the phrase 'survival of the fittest' see his letter to Wallace, 5 July 1866, MLD, vol. I, p. 270.

27 On the adoption of Spencer's philosophy by liberal Protestants see Moore, 'Herbert Spencer's Henchmen', 1985.

28 Spencer, *The Factors of Organic Evolution*, 1887, and 'The Inadequacy of Natural Selection', 1893. The latter was a response to Weismann's 'The All-Sufficiency of Natural Selection', 1893.

29 The story of the Darwin–Butler affair is told, with extracts from the relevant documents, by Nora Barlow in her edition of Darwin's *Autobiography*, part two, and Willey, *Darwin and Butler*, 1960.

30 Butler to Mivart, 29 February 1884, in H. F. Jones, *Samuel Butler: A Memoir*, 1919, vol. I, p. 407.

31 Butler, 'The Deadlock in Darwinism'; reprinted in Butler, *Essays*, 1908; see p. 308.

32 On the emergence of late nineteenth-century Lamarckism, see Bowler, *The Eclipse of Darwinism*, ch. 4.

33 See for instance Francis Darwin, 'President's Address', p. 14.

34 This pamphlet is reprinted in the account of the affair given in Darwin's *Autobiography*.

35 On the American school see Bowler, *The Eclipse of Darwinism*, ch. 6.

Chapter 10

1 Huxley did not say that he would rather be descended from an ape than from a bishop, but his reply was not very complimentary; see LLD, vol. II, pp. 320–3, and *Life and Letters of Huxley*, 1908, vol. I, pp. 259–74.

2 See Monypenny and Buckle, *The Life of Benjamin Disraeli*, 1929, vol. II, p. 108.

3 *Origin of Species* (1st edn), p. 488. On Darwin's reasons for inserting this sentence, see *Autobiography*, pp. 130–1.

4 See R. J. Richards, *Darwin and the Emergence of Evolutionary Theories of Mind and Behavior*, 1987.

5 See *Charles Darwin's Natural Selection*, 1975, ch. 10, *Origin of Species* (1st edn), ch. 7, and *Origin of Species* (6th edn), ch. 8.

6 See Desmond, *Archetypes and Ancestors*, 1982, ch. 2.

7 Darwin to Lyell, 6 March 1863, LLD, vol. III, p. 12. For the offending passage see Lyell, *Geological Evidences of the Antiquity of Man*, 1863, p. 505. See Bartholomew, 'Lyell and Evolution', 1973.

8 Darwin to Wallace, 14 April 1869, LLD, vol. III, p. 116. See Kottler, 'Alfred Russel Wallace, the Origin of Man, and Spiritualism', 1974.

9 LLD, vol. III, p.131.

10 *Descent of Man* (2nd edn, 1885), p. 103.

11 Ibid., pp. 49–53. In the two-volume first edition this passage occurs at vol. I, pp. 138–45 (the chapter 'On the Manner of Development of Man from Some Lower Form' was repositioned from chapter 4 to chapter 2 in the second edition). On the unique character of this insight compared with other nineteenth-century ideas on human origins see Bowler, *Theories of Human Evolution*, 1986, especially ch. 7.

12 *Descent of Man* (2nd edn), pp. 578–9. On biologists and the race question see Stepan, *The Idea of Race in Science*, 1982.

13 See Grayson, *The Establishment of Human Antiquity*, 1983.

14 On cultural evolutionism see Burrow, *Evolution and Society*, 1966, and Stocking, *Race, Culture, and Evolution*, 1968, and *Victorian Anthropology*, 1987.

15 Darwin to Lubbock, 11 June 1865, LLD, vol. III, p. 36.

16 See R. J. Richards, *Darwin and the Emergence of Evolutionary Theories of Mind and Behavior*, ch. 8.

17 Darwin to W. Graham, 3 July 1881, LLD, vol. I, p. 316.

18 See for instance Hofstadter, *Social Darwinism in American Thought*, 1959.

19 See Bannister, *Social Darwinism: Science and Myth in Anglo-American Social Thought*, 1979, and G. Jones, *Social Darwinism and English Thought*, 1980.

20 *Descent of Man* (2nd edn), p. 137.

21 To Lyell, 4 January 1860, LLD, vol. II, p. 262.

22 Huxley's 'Evolution and Ethics' is reprinted in vol. 9 of his *Collected Essays*, 1893–4.

23 See Spencer, *Principles of Biology*, 1864, vol. I, p. 444. On the differences between Darwin's and Spencer's use of the struggle concept see Bowler, *The Non-Darwinian Revolution*, 1988, pp. 38–40.

24 See Moore, 'Herbert Spencer's Henchmen', 1985.

25 See Forrest, *Francis Galton*, 1974, and on eugenics see Haller, *Eugenics*, 1963; Kevles, *In the Name of Eugenics*, 1985; and Searle, *Eugenics and Politics in Britain*, 1976.
26 Darwin to Galton, 4 January 1873, MLD, vol. II, p. 43. Hereditary genius is ⋅discussed in the *Descent of Man* (2nd edn), pp. 136–42.

Chapter 11

1 See LLD, vol. III, pp. 206–10.
2 Ibid., vol. III, pp. 351–4.
3 For a list of Darwin's honours see ibid., vol. III, appendix IV.
4 See Colp, 'The Contacts between Karl Marx and Charles Darwin', 1974; Fay, 'Did Marx Offer to Dedicate *Capital* to Darwin?', 1978; and Feuer, 'Is the Darwin–Marx Correspondence Authentic?', 1975.
5 See LLD, vol. I, p. 316.
6 Ibid., vol. I, p. 304.
7 To Wallace, July 1881, LLD, vol. III, p. 356.
8 Ibid., vol. III, p. 358.
9 On the funeral see ibid., vol. III, appendix I, and Moore, 'Charles Darwin Lies in Westminster Abbey', 1982.
10 See Bowler, *The Eclipse of Darwinism*, 1983, and *Evolution: The History of an Idea*, 1984, chs 9 and 11.
11 See Sulloway, 'Geographic Isolation in Darwin's Thinking', 1979.
12 J. Huxley, *Evolution: The Modern Synthesis*, 1942, pp. 22–8.
13 For a survey of current thinking on the origin of genetics, see Bowler, *The Mendelian Revolution*, 1989.
14 See Weismann, 'The All-Sufficiency of Natural Selection', 1893. On Weismann's theory of heredity see his *The Germ Plasm*, 1893. For a recent evaluation see Mayr, 'Weismann and Evolution', 1985.
15 See Provine, *The Origins of Theoretical Population Genetics*, 1971, and Mayr and Provine (eds), *The Evolutionary Synthesis*, 1980.
16 See Bowler, *The Invention of Progress*, 1989.
17 The controversy was sparked by Wilson's *Sociobiology: The New Synthesis*, 1975, and by later books such as his *On Human Nature*, 1978. The term 'selfish gene' was introduced by Richard Dawkins; see Dawkins, *The Selfish Gene*, 1976. For a survey of the controversy on the human implications of sociobiology see Caplan, *The Sociobiology Debate*, 1978.

Bibliography

Appel, T., *The Cuvier–Geoffroy Debate: French Biology in the Decades before Darwin*, New York and Oxford: Oxford University Press, 1987.

Argyll, G. D. Campbell, Duke of, *The Reign of Law*, London: Strahan, 1866.

Bannister, R. C., *Social Darwinism: Science and Myth in Anglo-American Social Thought*, Philadephia, PA: Temple University Press, 1979.

Bartholomew, M., 'Lyell and Evolution: An Account of Lyell's Response to the Prospect of an Evolutionary Ancestry for Man', *Br. J. Hist. Sci.,* 6 (1973): 261–303.

Barzun, J., *Darwin, Marx, Wagner: Critique of a Heritage*, 2nd edn, Garden City, NY: Doubleday, 1958.

Beddall, B. G., *Wallace and Bates in the Tropics: An Introduction to the Theory of Natural Selection*, London: Macmillan, 1969.

Bowler, P. J., 'Darwin's Concepts of Variation', *J. Hist. Medicine,* 29 (1974): 196–212.

—— *Fossils and Progress: Paleontology and the Idea of Progressive Evolution in the Nineteenth Century*, New York: Science History Publications, 1976.

—— 'Malthus, Darwin, and the Concept of Struggle', *J. Hist. Ideas,* 37 (1976): 631–50.

—— 'Alfred Russel Wallace's Concepts of Variation', *J. Hist. Medicine,* 31 (1976): 17–29.

—— 'Darwinism and the Argument from Design: Suggestions for a Re-evaluation', *J. Hist. Biology,* 10 (1977): 29–43.

—— *The Eclipse of Darwinism: Anti-Darwinian Evolution Theories in the Decades around 1900*, Baltimore, MD: Johns Hopkins University Press, 1983.

—— *Evolution: The History of an Idea*, Berkeley, CA: University of California Press, 1984 (new edn, 1989).

—— *Theories of Human Evolution: A Century of Debate, 1844–1944*, Baltimore, MD: Johns Hopkins University Press, and Oxford: Basil Blackwell, 1986.

Bowler, P. J., *The Non-Darwinian Revolution: Reinterpreting a Historical Myth*, Baltimore, MD: Johns Hopkins University Press, 1988.

—— *The Mendelian Revolution: The Emergence of Hereditarian Concepts in Modern Science and Society*, London: Athlone Press, 1989.

—— *The Invention of Progress: The Victorians and the Past*, Oxford: Basil Blackwell, 1989.

Brackman, A., *A Delicate Arrangement: The Strange Case of Charles Darwin and Alfred Russel Wallace*, New York: Times Books, 1980.

Brooks, J. L., *Just Before the Origin: Alfred Russel Wallace's Theory of Evolution*, New York: Columbia University Press, 1983.

Brown, F. B., 'The Evolution of Darwin's Theism', *J. Hist. Biology*, 19 (1986): 1–45.

Browne, J., 'Darwin's Botanical Arithmetic and the "Principle of Divergence", 1854–1858', *J. Hist. Biololgy*, 13 (1980): 53–89.

—— *The Secular Ark: Studies in the History of Biogeography*, New Haven, CT: Yale University Press, 1983.

Buckland, W., *Reliquiae Diluvianae: or Observations on the Organic Remains Contained in Caves ... Attesting the Action of a Universal Deluge*, 2nd edn, London: John Murray, 1824.

—— *Bridgewater Treatise: Geology and Mineralogy Considered with Reference to Natural Theology*, 2nd edn, London, 1837, 2 vols.

Buffetaut, E., *A Short History of Vertebrate Palaeontology*, London: Croom Helm, 1987.

Burchfield, J. D., 'Darwin and the Dilemma of Geological Time', *Isis*, 65 (1974): 301–21.

—— *Lord Kelvin and the Age of the Earth*, New York: Science History Publications, 1975.

Burkhardt, R. W., Jr., *The Spirit of System: Lamarck and Evolutionary Biology*, Cambridge, MA: Harvard University Press, 1977.

Burrow, J. W., *Evolution and Society: A Study in Victorian Social Theory*, Cambridge: Cambridge University Press, 1966.

Butler, S., *Evolution, Old and New*, London: Harwicke and Bogue, 1879.

—— *Life and Habit*, new edn., London: A. C. Fifield, 1916.

—— *Essays on Life, Art and Science* [1908], reprinted Port Washington, NY: Kennikat Press, 1970.

Cannon, W. F., 'The Bases of Darwin's Achievement: a Revaluation', *Victorian Stud.*, 5 (1961): 109–32.

Caplan, A. L. (ed.), *The Sociobiology Debate*, New York: Harper and Row, 1978.

Carpenter, W. B., *Nature and Man: Essays Scientific and Philosophical*, London: Kegan Paul, Trench, 1888.

Chambers, R., *Vestiges of the Natural History of Creation*, London: John Churchill, 1844 (5th edn, 1846; 11th edn, 1860).

Clark, R. W., *The Survival of Charles Darwin*, London: Weidenfeld and Nicolson, 1985.

Colp, R., Jr., 'The Contacts between Karl Marx and Charles Darwin', *J. Hist. Ideas*, 35 (1974): 329–38.

—— *To be an Invalid: the Illness of Charles Darwin*, Chicago, IL: University of Chicago Press, 1977.

Combe, G., *Of the Constitution of Man*, Edinburgh, 1828.

Cooter, R., *The Cultural Meaning of Popular Science: Phrenology and the Organization of Consent in Nineteenth-century Britain*, Cambridge: Cambridge University Press, 1985.

Corsi, P., *Science and Religion: Baden Powell and the Anglican Debate, 1800–1860*, Cambridge: Cambridge University Press, 1988.

—— *The Age of Lamarck*, Berkeley, CA: University of California Press, 1988.

Cuvier, G., *An Essay on the Theory of the Earth*, Edinburgh, 1813.

—— *Recherches sur les ossements fossiles*, 3rd edn, Paris: Dufour et D'Ocagne, 1825, 5 vols.

Darwin, C. R. (ed.), *The Zoology of the Voyage of HMS Beagle*, London, 1838–43, 5 parts.

—— *The Structure and Distribution of Coral Reefs*, London, 1842.

—— *Geological Observations on the Volcanic Islands Visited during the Voyage of HMS Beagle*, London, 1844.

—— *Journal of Researches into the Geology and Natural History of the Countries Visited during the Voyage of HMS Beagle* [1845], reprinted London: Routledge, 1891.

—— *Geological Observations on South America*, London, 1846.

—— *A Monograph of the Sub-class Cirripedia with Figures of all the Species*, London: Ray Society, 1851–4, 2 vols.

—— *A Monograph of the Fossils Lepadidae*, London: Palaeontographical Society, 1851–4, 2 vols.

—— *On the Origin of Species by Means of Natural Selection: or the Preservation of Favoured Races in the Struggle for Life*, London, 1859, reprinted Cambridge, MA: Harvard University Press, 1964.

—— *On the Various Contrivances by which British and Foreign Orchids are Fertilized by Insects*, London: John Murray, 1862.

—— *The Movements and Habits of Climbing Plants*, London: John Murray, 1865.

—— *The Variation of Animals and Plants under Domestication*, London: John Murray, 1868, 2 vols (2nd edn, 1882, 2 vols).

—— *The Descent of Man and Selection in Relation to Sex*, London: John Murray, 1871, 2 vols (2nd edn, revised, 1885).

—— *The Expression of the Emotions in Man and the Animals*, London: John Murray, 1872, 2 vols.

—— *Insectivorous Plants*, London: John Murray, 1875.

—— *The Effect of Cross and Self-Fertilization in the Vegetable Kingdom*, London: John Murray, 1876.

—— *The Different Forms of Flowers on Plants of the Same Species*, London: John Murray, 1877.

—— *The Power of Movement in Plants*, London: John Murray, 1880.

—— *The Formation of Vegetable Mould through the Action of Worms*, London: John Murray, 1881.

—— *The Life and Letters of Charles Darwin*, ed. by F. Darwin, London: John Murray, 1887, 3 vols.

—— *More Letters of Charles Darwin*, ed. by F. Darwin, London: John Murray, 1903, 2 vols.

Darwin, C. R., *Charles Darwin's Diary of the Voyage of HMS Beagle*, ed. by N. Barlow, 1933, reprinted New York: Krause Reprints, 1969.
—— *Charles Darwin and the Voyage of the Beagle: Unpublished Letters and Notebooks*, ed. by N. Barlow, London: Pilot Press, 1945.
—— *The Autobiography of Charles Darwin*, ed. by N. Barlow, New York: Harcourt, Brace, 1958.
—— *The Origin of Species . . . a Variorum Text*, ed. by M. Peckham, Philadelphia, PA: University of Pennsylvania Press, 1959.
—— *Charles Darwin's Natural Selection: being the Second Part of his Big Species Book*, ed. by R. C. Stauffer, Cambridge: Cambridge University Press, 1975.
—— *The Collected Papers of Charles Darwin*, ed. by P. H. Barrett, Chicago, IL: University of Chicago Press, 1977, 2 vols.
—— *A Concordance to Darwin's Origin of Species, First Edition*, ed. by P. H. Barrett, D. J. Weinshank and T. T. Gottleber, Ithaca, NY: Cornell University Press, 1981.
—— *A Calendar of the Correspondence of Charles Darwin, 1821–1882*, ed. by F. Burkhardt and S. Smith, New York: Garland Publishing, 1985.
—— *The Correspondence of Charles Darwin*, vol. I (1821–36), vol. II (1837–43), vol. III (1844–6), Cambridge: Cambridge University Press, 1985–8.
—— *Charles Darwin's Notebooks, 1836–1844*, ed. by P. H. Barrett, D. J. Weinshank and T. T. Gottleber, London: British Museum (Natural History) and Cambridge: Cambridge University Press, 1987.
—— *Charles Darwin's Beagle Diary*, ed. by R. D. Keynes, Cambridge: Cambridge University Press, 1988.
Darwin, C., and Wallace, A. R., *Evolution by Natural Selection*, Introduction by Sir Gavin De Beer, Cambridge: Cambridge University Press, 1958.
Darwin, E., *The Botanic Garden*, London, 1791.
—— *Zoonomia, or the Laws of Organic Life*, London, 1794–6, 2 vols.
—— *The Temple of Nature*, London, 1803.
—— *The Essential Erasmus Darwin*, ed. by D. King-Hele, London: McGibbon and Kee, 1968.
—— *The Letters of Erasmus Darwin*, ed. by D. King-Hele, Cambridge: Cambridge University Press, 1981.
Darwin, F., 'President's Address', *Report of the British Association for the Advancement of Science,* 1908 meeting, pp. 3–27.
Dawkins, R., *The Selfish Gene*, Oxford: Oxford University Press, 1976.
De Beer, Sir G., *Charles Darwin*, London: Nelson, 1963.
Desmond, A., 'Designing the Dinosaurs: Richard Owen's Response to Robert Edmund Grant', *Isis*, 70 (1979): 224–34.
—— *Archetypes and Ancestors: Palaeontology in Victorian London 1850–1875*, London: Blond and Briggs, 1982.
—— 'Robert E. Grant: The Social Predicament of a Pre-Darwinian Evolutionist', *Br. J. Hist. Sci.*, 17 (1984): 189–223.
—— 'Richard Owen's Reaction to Transmutation in the 1830s', *Br. J. Hist. Sci.*, 18 (1985): 25–50.

——'Artisan Resistance and Evolution in Britain, 1819–1848', *Osiris,* 2nd series, 3 (1987): 77–110.

—— *The Politics of Evolution: Morphology, Medicine, and Reform in Radical London,* Chicago, IL: University of Chicago Press, 1989.

Di Gregorio, M., *T. H. Huxley's Place in Natural Science,* New Haven, CT: Yale University Press, 1984.

Dobshansky, T., *Genetics and the Origin of Species,* New York: Columbia University Press, 1937.

Drummond, H., *The Ascent of Man,* New York: James Pott, 1894.

Dupree, A. H., *Asa Gray,* Cambridge, MA: Harvard University Press, 1959.

Durant, J. (ed.), *Darwinism and Divinity: Essays on Evolution and Religious Belief,* Oxford: Basil Blackwell, 1985.

Egerton, F. N., 'Refutation and Conjecture: Darwin's Response to Sedgwick's Attack on Chambers', *Stud. Hist. Phil. Sci.,* 1 (1970): 176–83.

Eiseley, L., *Darwin's Century: Evolution and the Men Who Discovered It,* New York: Doubleday, 1958.

Ellegard, A., *Darwin and the General Reader: The Reception of Darwin's Theory of Evolution in the British Periodical Press, 1859–1872,* Goteburg: Acta Universitatis Gothenburgensis, 1957.

Fay, M., 'Did Marx Offer to Dedicate *Capital* to Darwin?', *J. Hist. Ideas,* 39 (1978): 133–46.

Feuer, L. S., 'Is the Darwin–Marx Correspondence Authentic?', *Annals of Science,* 32 (1975): 1–12.

Fichman, M., *Alfred Russel Wallace,* Boston, MA: Twayne, 1981.

Fisher, R. A., *The Genetical Theory of Natural Selection,* Oxford: Clarendon Press, 1930.

Forrest, D., *Francis Galton: The Life and Work of a Victorian Genius,* New York: Tapplinger, 1974.

Galton, F., *Hereditary Genius,* London: Macmillan, 1869.

Geison, G., 'Darwin and Heredity: the Evolution of his Hypothesis of Pangenesis', *J. Hist. Medicine,* 24 (1969): 375–411.

George, W., *Biologist–Philosopher: A Study of the Life and Writings of Alfred Russel Wallace,* New York: Abelard-Schumann, 1964.

Ghiselin, M. T., *The Triumph of the Darwinian Method,* Berkeley, CA: University of California Press, 1969.

Gillispie, C. C., *Genesis and Geology: A Study in the Relations of Scientific Thought, Natural Theology, and Social Opinion in Great Britain, 1790–1850,* Cambridge, MA: Harvard University Press, 1951.

[Grant, R. E.], 'Observations on the Nature and Importance of Geology', *Edinburgh New Phil. J.,* 1 (1826): 293–302.

——'Of the Changes which Life has Experienced on the Globe', *Edinburgh New Phil. J.,* 3 (1827): 298–301.

Gray, A., *Darwiniana: Essays and Reviews Pertaining to Darwinism,* New York: Appleton, 1876.

Grayson, D. K., *The Establishment of Human Antiquity,* New York: Academic Press, 1983.

Greene, J. C., *The Death of Adam: Evolution and its Impact on Western Thought*, Ames, IA: Iowa State University Press, 1959.

—— *Science, Ideology and World View*, Berkeley, CA: University of California Press, 1981.

Gruber, H. E., *Darwin on Man: a Psychological Study of Scientific Creativity*, New York: E. P. Dutton, 1974.

Gruber, J. W., *A Conscience in Conflict: The Life of St. George Jackson Mivart*, New York: Columbia University Press, 1960.

Haeckel, E., *The History of Creation: Or the Development of the Earth and its Inhabitants by the Action of Natural Causes. A Popular Exposition of the Doctrine of Evolution in General and of that of Darwin, Lamarck, and Goethe in Particular*, New York: Appleton, 1876, 2 vols.

—— *The Evolution of Man: A Popular Exposition of the Principal Points of Human Ontogeny and Phylogeny*, New York: Appleton, 1879, 2 vols.

—— *The Last Link: Our Present Knowledge of the Descent of Man*, London: A. &. C. Black, 1898.

Haller, M., *Eugenics: Hereditarian Attitudes in American Thought*, New Brunswick, NJ: Rutgers University Press, 1963.

Herbert, S., 'Darwin, Malthus, and Selection', *J. Hist. Biology*, 4 (1971): 209–17.

—— 'The Place of Man in the Development of Darwin's Theory', *J. Hist. Biology*, 7 (1974): 217–58; 10 (1977): 155–227.

—— 'Darwin the Young Geologist'. In D. Kohn (ed.), *The Darwinian Heritage*, Princeton, NJ: Princeton University Press, 1985, pp. 483–510.

Herschel, Sir J. F. W., *A Preliminary Discourse on the Study of Natural Philosophy*, London, 1831.

—— *Physical Geography*, Edinburgh: A. & C. Black, 1861.

Himmelfarb, G., *Darwin and the Darwinian Revolution*, New York: Norton, 1959.

Hodge, M. J. S., 'Lamarck's Science of Living Bodies', *Br. J. Hist. Sci.*, 5 (1971): 323–52.

—— 'The Universal Gestation of Nature: Chambers' *Vestiges* and *Explanations*', *J. Hist. Biology*, 5 (1972): 127–52.

—— 'Darwin and the Laws of the Animate Part of the Terrestrial System (1835–1837): On the Lyellian Origins of his Zoonomical Explanatory Programme', *Stud. Hist. Biology*, 6 (1982): 1–106.

—— 'Darwin as a Lifelong Generation Theorist'. In D. Kohn (ed.), *The Darwinian Heritage*, Princeton, NJ: Princeton University Press, 1985, pp. 207–44.

Hodge, M. J. S., and Kohn, D., 'The Immediate Origins of Natural Selection'. In D. Kohn (ed.), *The Darwinian Heritage*, Princeton, NJ: Princeton University Press, 1985, pp. 185–206.

Hofstadter, R., *Social Darwinism in American Thought*, revised edition, Boston, MA: Beacon Press, 1959.

Hooker, J. D., 'On the Origination and Distribution of Vegetable Species: Introductory Essay to the Flora of Tasmania', *Am. J. Sci., 2nd series*, 29 (1860): 1–25, 305–26.

—— *The Life and Letters of Sir Joseph Dalton Hooker*, ed. by L. Huxley, London, 1918, 2 vols.

BIBLIOGRAPHY

Hooykaas, R., *Natural Law and Divine Miracle: the Principle of Uniformity in Geology, Biology, and History*, Leiden: Brill, 1959.

Hull, D. L. (ed.), *Darwin and his Critics: the Reception of Darwin's Theory of Evolution by the Scientific Community*, Cambridge, MA: Harvard University Press, 1973.

—— 'Darwinism as a Historical Entity: a Historiographical Proposal'. In D. Kohn (ed.), *The Darwinian Heritage*, Princeton, NJ: Princeton University Press, 1985, pp. 773–812.

Humboldt, A. von, and Bonpland, A., *Personal Narrative of Travels to the Equinoctal Regions of the New Continent during the Years 1799–1804*, London, 1814–29, 7 vols.

Huxley, J., *Evolution: The Modern Synthesis*, London: Allen and Unwin, 1942.

Huxley, T. H., 'Vestiges of the Natural History of Creation', *Br. Foreign Med. Chirurg. Rev.*, 13 (1853): 332–43.

—— 'On the Reception of the Origin of Species'. In C. R. Darwin, *The Life and Letters of Charles Darwin*, ed. by F. Darwin, London: John Murray, 1887, vol. II, pp. 179–204.

—— *Man's Place in Nature*, London: Williams and Norgate, 1863.

—— *American Addresses: With a Lecture on the Study of Biology*, New York: Appleton, 1888.

—— *Collected Essays*, London: Macmillan, 1893–4, 9 vols (vol. 2, *Darwiniana*; vol. 4, *Science and Hebrew Tradition*; vol. 7, *Man's Place in Nature*; vol. 9, *Evolution and Ethics*).

—— *The Life and Letters of Thomas Henry Huxley*, ed by L. Huxley, London: Macmillan, 1908, 3 vols.

Jenkin, F., 'The Origin of Species', *North British Review*, 46 (1867): 277–318.

Jensen, J. V., 'Return to the Huxley–Wilberforce Debate', *Br. J. Hist. Sci.*, 21 (1988): 161–80.

Jones, G., *Social Darwinism and English Thought*, London: Harvester, 1980.

Jones, H. F., *Samuel Butler, Author of Erewhon (1835–1902): A Memoir*, London: Macmillan, 1919, 2 vols.

Jordanova, L., *Lamarck*, Oxford: Oxford University Press, 1984.

Kevles, D., *In the Name of Eugenics: Genetics and the Uses of Human Heredity*, New York: Knopf, 1985.

Keynes, R. D., *The Beagle Record: Selections from the Original Pictorial Records and Written Accounts of the Voyage of HMS Beagle*, Cambridge: Cambridge University Press, 1979.

King-Hele, D., *Erasmus Darwin*, New York: Scribner, 1963.

Koestler, A., *The Ghost in the Machine*, New York: Macmillan, 1967.

—— *The Case of the Midwife Toad*, London: Hutchinson, 1971.

Kohn, D. (ed.), *The Darwinian Heritage*, Princeton, NJ: Princeton University Press, 1985.

—— 'Darwin's Principle of Divergence as Internal Dialogue'. In D. Kohn (ed.), *The Darwinian Heritage*, Princeton, NJ: Princeton University Press, 1985, pp. 245–57.

—— 'Darwin's Ambiguity: the Secularization of Biological Meaning', *Br. J. Hist. Sci.*, 22 (1989): 215–40.

Kottler, M. J., 'Alfred Russel Wallace, the Origin of Man, and Spiritualism', *Isis*, 65 (1974): 145–92.

Kottler, M. J., 'Charles Darwin and Alfred Russel Wallace: Two Decades of Debate over Natural Selection'. In D. Kohn (ed.), *The Darwinian Heritage*, Princeton, NJ: Princeton University Press, 1985, pp. 367–432.

Lamarck, J. B., *Zoological Philosophy*, trans. by H. Elliot, London, 1914 (reprinted New York: Hafner, 1963).

Lovtrup, S., *Darwinism: the Refutation of a Myth*, London: Croom Helm, 1987.

Lubbock, J., *Prehistoric Times: As Illustrated by Ancient Remains and the Manners and Customs of Modern Savages*, London: Williams and Norgate, 1865.

Lyell, C., *Principles of Geology*, London, 1830–3, 3 vols (7th edn, London: John Murray, 1847).

—— *Geological Evidences of the Antiquity of Man*, London: John Murray, 1863.

McKinney, H. L., *Wallace and Natural Selection*, New Haven, CT: Yale University Press, 1972.

MacLeod, R., 'The X-Club: a Scientific Network in Late-Victorian England', *Notes and Records Roy. Soc. Lond.*, 24 (1970): 305–22.

McNeil, M., *Under the Banner of Science: Erasmus Darwin and his Age*, Manchester: Manchester University Press, 1987.

Mayr, E., *Systematics and the Origin of Species*, New York: Columbia University Press, 1942.

—— *The Growth of Biological Thought*, Cambridge, MA: Harvard University Press, 1982.

—— 'Weismann and Evolution', *J. Hist. Biology*, 18 (1985): 259–322.

Mayr, E., and Provine, W. B. (eds), *The Evolutionary Synthesis: Perspectives on the Unification of Biology*, Cambridge, MA: Harvard University Press, 1980.

Miller, H., *The Old Red Sandstone: Or New Walks in an Old Field*, Edinburgh, 1841 (new edn, Boston, 1858).

—— *Footprints of the Creator: Or the Asterolepis of Stromness*, 3rd edn, London: Johnstone and Hunter, 1850.

Millhauser, M., *Just before Darwin: Robert Chambers and Vestiges*, Middletown, CT: Wesleyan University Press, 1959.

Mivart, St. G. J., *On the Genesis of Species*, London: Macmillan, 1871.

—— *Man and Apes: An Exposition of Structural Resemblances Bearing upon Questions of Affinity and Origin*, London: Robert Harwicke, 1873.

Monypenny, W. F., and Buckle, G. E., *The Life of Benjamin Disraeli*, revised edition, London: John Murray, 1929, 2 vols.

Moore, J. R., *The Post-Darwinian Controversies: A Study of the Protestant Struggle to Come to Terms with Darwin in Britain and America, 1870–1900*, Cambridge: Cambridge University Press, 1979.

—— 'Charles Darwin Lies in Westminster Abbey', *Biol. J. Linnean Soc.*, 17 (1982): 97–113.

—— 'Herbert Spencer's Henchmen: The Evolution of Protestant Liberals in Late-Nineteenth-Century America'. In J. Durant (ed.), *Darwinism and Divinity: Essays on Evolution and Religious Belief*, Oxford: Basil Blackwell, 1985, pp. 76–100.

—— 'Darwin of Down: The Evolutionist as Squarson–Naturalist'. In D. Kohn (ed.), *The Darwinian Heritage*, Princeton, NJ: Princeton University Press, 1985, pp. 435–82.

——(ed.), *History, Humanity, and Evolution: Essays in Honour of John C. Greene*, Cambridge: Cambridge University Press, 1989.

Moorehead, A., *Darwin and the Beagle*, London: Hamish Hamilton, 1969.

Nelkin, D., *The Creation Controversy: Science or Scripture in the Public Schools*, New York: Norton, 1983.

Nicholson, A. J., 'The Role of Population Dynamics in Natural Selection'. In S. Tax (ed.), *Evolution after Darwin*, Chicago, IL: University of Chicago Press, 1960, 3 vols, vol I. pp. 477–522.

Olby, R. C., 'Charles Darwin's Manuscript of Pangenesis', *Br. J. Hist., Sci.* 1 (1963): 251–63.

Oldroyd, D. R., 'How did Darwin Arrive at his Theory?', *History of Science*, 22 (1984): 325–74.

Ospovat, D., 'The Influence of Karl Ernst von Baer's Embryology, 1828–1859: A Reappraisal in Light of Richard Owen and William B. Carpenter's "Paleontological Application of von Baer's Law" ', *J. Hist. Biology*, 9 (1976): 1–28.

—— *The Development of Darwin's Theory. Natural History, Natural Theology and Natural Selection, 1838–1859*, Cambridge: Cambridge University Press, 1981.

Owen, R., 'Report on British Fossil Reptiles, Part 2', *Report of the British Association for the Advancement of Science*, 1841, pp. 60–204.

—— *On the Nature of Limbs*, London: van Voorst, 1849.

—— 'On the Origin of Species', *Edinburgh Review*, 111 (1860): 487–532.

—— *On the Anatomy of the Vertebrates*, London: Longmans, Green, 1866–8, 3 vols.

Paley, W., *Natural Theology: or Evidences of the Existence and Attributes of the Deity Collected from the Appearances of Nature*, London, 1802.

Paradis, J. G., *T. H. Huxley: Man's Place in Nature*, Lincoln, NE: University of Nebraska Press, 1978.

Peel, J. D. Y., *Herbert Spencer: The Evolution of a Sociologist*, London: Heinemann, 1971.

Porter, D. M., 'The *Beagle* Collector and his Collections'. In D. Kohn (ed.), *The Darwinian Heritage*, Princeton, NJ: Princeton University Press, 1985, pp. 973–1019.

Poulton, E. B., *Charles Darwin and the Origin of Species*, London: Longmans, Green, 1909.

Powell, B., *Essays on the Spirit of the Inductive Philosophy, The Unity of Worlds, and the Philosophy of Creation*, London: 1855.

Provine, W. B., *The Origins of Theoretical Population Genetics*, Chicago, IL: University of Chicago Press, 1971.

Rachootin, S., 'Owen and Darwin Reading a Fossil: *Macraucheria* in a Boney Light'. In D. Kohn (ed.), *The Darwinian Heritage*, Princeton, NJ: Princeton University Press, 1985, pp. 155–83.

Ralling, C., *The Voyage of Charles Darwin*, London: BBC, 1978.

Rehbock, P. F., *The Philosophical Naturalists: Themes in Early Nineteenth-Century British Biology*, Madison, WI: University of Wisconsin Press, 1983.

Richards, E., 'A Question of Property Rights: Richard Owen's Evolutionism Reassessed', *Br. J. Hist. Sci.*, 20 (1987): 129–72.

Richards, R. J., *Darwin and the Emergence of Evolutionary Theories of Mind and Behavior*, Chicago, IL: University of Chicago Press, 1987.

Romanes, G. J., *Animal Intelligence*, London: Kegan Paul, Trench, 1881.

—— *Mental Evolution in Animals*, London: Kegan Paul, Trench, 1883.

—— *Mental Evolution in Man: Origins of Human Faculty*, London: Kegan Paul, Trench, Trübner, 1888.

Rudwick, M. J. S., 'Uniformity and Progression: Reflections on the Structure of Geological Theory in the Age of Lyell'. In D. H. D. Roller (ed.), *Perspectives in the History of Science and Technology*, Norman, OK: University of Oklahoma Press, 1971, pp. 209–27.

—— *The Meaning of Fossils: Episodes in the History of Palaeontology*, New York: American Elsevier, 1972.

—— 'Darwin and Glen Roy: a "Great Failure" in Scientific Method?', *Stud. Hist. Phil. Sci.*, 5 (1974): 97–185.

—— 'Charles Darwin in London: The Integration of Public and Private Science', *Isis*, 73 (1982): 186–206.

—— *The Great Devonian Controversy: The Shaping of Scientific Knowledge among Gentlemanly Specialists*, Chicago, IL: University of Chicago Press, 1985.

—— 'A Year in the Life of Adam Sedgwick and Company, Geologists', *Archives of Natural History*, 15 (1988): 243–68.

Rupke, N., *The Great Chain of History: William Buckland and the English School of Geology, 1814–1849*, Oxford: Clarendon Press, 1983.

Ruse, M., 'Darwin's Debt to Philosophy: an Examination of the Influence of the Philosophical Ideas of John F. W. Herschel and William Whewell', *Stud. Hist. Phil. Sci.*, 6 (1975): 159–81.

—— *The Darwinian Revolution: Science Red in Tooth and Claw*, Chicago, IL: University of Chicago Press, 1979.

Schweber, S. S., 'The Origin of the *Origin* Revisited', *J. Hist. Biology*, 10 (1977): 229–316.

—— 'Darwin and the Political Economists: Divergence of Character', *J. Hist. Biology*, 13 (1980): 195–289.

—— 'The Wider British Context in Darwin's Theorizing'. In D. Kohn, (ed.), *The Darwinian Heritage*, Princeton, NJ: Princeton University Press, 1985, pp. 35–69.

—— 'John Herschel and Charles Darwin: A Study in Parallel Lives', *J. Hist. Biology*, 22 (1989): 1–71.

Searle, G. R., *Eugenics and Politics in Britain, 1900–1914*, Leiden: Noordhoff, 1976.

Secord, J. A., 'Darwin and the Breeders: a Social History'. In D. Kohn (ed.), *The Darwinian Heritage*, Princeton, NJ: Princeton University Press, 1985, pp. 519–42.

—— *Controversy in Victorian Geology: The Cambrian–Silurian Dispute*, Princeton, NJ: Princeton University Press, 1986.

—— 'Behind the Veil: Robert Chambers and the Genesis of the *Vestiges of Creation*'. In J. R. Moore (ed.), *History, Humanity, and Evolution*, Cambridge: Cambridge University Press, 1989, pp. 165–94.

Sedgwick, A., 'Vestiges of the Natural History of Creation', *Edinburgh Rev.*, 82 (1845): 1–85.

——— *The Life and Letters of the Reverend Adam Sedgwick*, ed. by J. W. Clark and T. M. Hughes, Cambridge: Cambridge University Press, 1890, 2 vols.

Simpson, G. G., *Tempo and Mode in Evolution*, New York: Columbia University Press, 1944.

Sloan, P. R., 'Darwin's Invertebrate Program, 1826–1836: Preconditions for Transformism'. In D. Kohn (ed.), *The Darwinian Heritage*, Princeton, NJ: Princeton University Press, 1985, pp. 71–120.

——— 'Darwin, Vital Matter, and the Transformism of Species', *J. Hist. Biology*, 19 (1986): 369–445.

Spencer, H., *Social Statics: Or the Conditions Essential to Human Happiness Specified, and the First of them Developed*, London: John Chapman, 1851.

——— *First Principles of a New Philosophy*, London: Williams and Norgate, 1861.

——— *Principles of Biology*, London: Williams and Norgate, 1864, 2 vols.

——— *Principles of Psychology*, 2nd edn, London: Williams and Norgate, 1870–2, 2 vols.

——— *Essays, Scientific, Political, and Speculative*, London: Williams and Norgate, 1883, 3 vols.

——— *The Factors of Organic Evolution*, London: Williams and Norgate, 1887.

——— 'The Inadequacy of Natural Selection', *Contemporary Review*, 43 (1893): 153–66, 439–56.

Stepan, N., *The Idea of Race in Science: Great Britain, 1800–1960*, London: Macmillan, 1982.

Stocking, G. W., Jr., *Race, Culture, and Evolution: Essays in the History of Anthropology*, New York: Free Press, and London: Collier-Macmillan, 1968.

——— *Victorian Anthropology*, New York: Free Press, 1987.

Sulloway, F., 'Geographic Isolation in Darwin's Thinking: the Vicissitudes of a Crucial Idea', *Stud. Hist. Biology*, 3 (1979): 23–65.

——— 'Darwin and his Finches: the Evolution of a Legend', *J. Hist. Biology*, 15 (1982): 1–54.

——— 'Darwin's Conversion: the *Beagle* Voyage and its Aftermath', *J. Hist. Biology*, 15 (1982): 325–96.

——— 'Darwin's Early Intellectual Development: an Overview of the *Beagle* Voyage (1831–1836)', In D. Kohn (ed.), *The Darwinian Heritage*, Princeton, NJ: Princeton University Press, 1985, pp. 121–54.

Turner, F. M., *Between Science and Religion: The Reaction to Scientific Naturalism in Late Victorian England*, New Haven, CT: Yale University Press, 1974.

Tylor, E. B., *Researches into the Early History of Mankind and the Development of Civilization*, 2nd edn, London: John Murray, 1870.

——— *Anthropology: An Introduction to the Study of Man and Civilization*, London: Macmillan, 1881.

Vorzimmer, P. J., *Charles Darwin: The Years of Controversy*, Philadelphia, PA: Temple University Press, 1970.

Wallace, A. R., *Contributions to the Theory of Natural Selection*, London: Macmillan, 1870.

——— *Darwinism: An Exposition of the Theory of Natural Selection*, London: Macmillan, 1889.

Wallace, A. R., *Natural Selection and Tropical Nature*, new edn, London: Macmillan, 1895.
—— *My Life: A Record of Events and Opinions*, London: Macmillan, 1905.
—— *Alfred Russel Wallace: Letters and Reminiscences*, ed. by A. Marchant, New York: Harper, 1916.
Weismann, A., 'The All-Sufficiency of Natural Selection', *Contemporary Review*, 64 (1893): 309–38, 596–610.
—— *The Germ Plasm: A Theory of Heredity*, London: Scott, 1893.
Whewell, W., *History of the Inductive Sciences*, 2nd edn, London, 1847, 3 vols.
—— *Philosophy of the Inductive Sciences*, 2nd edn, London, 1847, 2 vols.
White, A. D., *A History of the Warfare of Science with Theology in Christendom*, 1896 (reprinted New York: Dover, 1960), 2 vols.
Willey, B., *Darwin and Butler: Two Versions of Evolution*, London: Chatto and Windus, 1960.
Williams-Ellis, A., *Darwin's Moon: A Biography of Alfred Russel Wallace*, London: Blackie, 1958.
Wilson, E. O., *Sociobiology: The New Synthesis*, Cambridge, MA: Harvard University Press, 1975.
—— *On Human Nature*, Cambridge, MA: Harvard University Press, 1978.
Young, R. M. 'Malthus and the Evolutionists: the Common Context of Biological and Social Theory', *Past and Present*, 43 (1969): 109–45.
—— 'Darwin's Metaphor: Does Nature Select?', *Monist*, 55 (1971): 442–503.
—— *Darwin's Metaphor: Nature's Place in Victorian Culture*, Cambridge: Cambridge University Press, 1985.

Index

PRINTED IN GREAT BRITAIN